A Primer for Environmental Literacy

A Primer for Environmental Literacy

FRANK B. GOLLEY

Yale University Press   New Haven and London

Published with assistance from the Louis Stern Memorial Fund.

Designed by James J. Johnson and set in Melior and Gill Sans types by Running Feet Books, Durham, North Carolina.
Printed in the United States of America by Edwards Brothers, Ann Arbor, Michigan.

Library of Congress Cataloging-in-Publication Data

Golley, Frank B.
  A primer for environmental literacy / Frank B. Golley.
    p.   cm.
  Includes bibliographical references.
  ISBN 0-300-07315-1 (cloth : alk. paper). — ISBN 0-300-07049-7 (pb : alk. paper)
  1. Environmental sciences—Study and teaching.   2. Environmental education.
  I. Title.
  GE70.G67   1998
  577—dc21   97-34440

A catalogue record for this book is available from the British Library.

The paper in this book meets the guidelines for permanence and durability of the Committee on Production Guidelines for Book Longevity of the Council on Library Resources.

10 9 8 7 6 5 4 3 2 1

Contents

Cluster Three: The Population and the Individual

Cluster Four: Interaction Between Individuals and Species

Preface

This book is based on a core course developed for the graduate certificate program in environmental ethics at the University of Georgia. Its purpose is to present key concepts of environmental science in a concise format that can be understood by those who are not natural scientists. Mathematical and chemical equations are kept to a minimum, and the models are mainly conceptual and diagrammatic. The book should be of use in a broad range of courses and to professionals concerned about the environment, environmental ethics, and environmental literacy.

Although this volume can be used as a textbook, its primary purpose is to present a way of thinking that integrates what we know with how we value the environment. It is designed to supplement the many books on environmental ethics written from a philosophical perspective. The discussion of a concept usually touches on its implications for ethical value systems and environmentalism. These discussions are not meant to be fully developed arguments about environmental ethics. Rather, they are designed to provide a bridge between science, philosophy, and management. Readers may want to expand their understanding of environmental ethics by consulting some of the books and papers cited at the end of each chapter.

One of the major difficulties I have had in teaching the course has been accommodating the widely different backgrounds of the students. Some have had no science training since high school, while others have a wide experience in environmental science. This problem is handled is through a "T-education" approach to the topic, which is both broad (the arm of the T) and deep (the stem of the T). Breadth is provided by this book and additional readings. Depth is created by having each student read and research a topic of personal interest, taking part in weekly field exercises, and preparing research papers that are presented in oral and written form.

Readers seeking a deeper discussion of environmental concepts should refer to conventional ecology or environmental science texts, of which there are many, or to the primary literature. A list of readings in ecology and environmental ethics is given for each concept. These lists are not meant to be exhaustive, nor do they always include the most up-to-date articles and books. But they are enough to lead energetic students to key words and bibliographies that will open up a cornucopia of information. In the jargon of the backpacker, the goal is to "go light"—that is, to pack no more than is necessary for any experience one may encounter.

Many concepts are illustrated by one or more examples from the literature of environmental science. Reviewers of the manuscript asked for even more examples, as is conventional to attract readers' interest and to present a broader array of information. I view environmental science as a field-oriented science; illustrations of concepts are developed by field work, not through reading indoors or using a computer. For this reason, I encourage readers to use their imagination in devising ways to illustrate the concepts of environmental science by reference to plants, animals, and habitats in the field, forest, lake, and river. Even the casual reader should seek examples in the real world of backyard, city park, or countryside.

My courses always end with a critical review by the students, and the organization of the book reflects their reactions. I am particularly grateful to Lee Carrubba, Susan Fisher, Dawn Garrison, Toshihide Hamazaki, Coeli Hoover, Lena Lane, Hlynur Oskarsson, and William Straw for their contributions. I thank the colleagues who reviewed sections of the book, including Betty Jean Craig, Victoria Davion, Frederick Ferré, Patricia Gowaty, Jean Marie Hartman, Eugene Odum, and Holmes Ralston. Priscilla Golley prepared the figures and tables; I thank her for this essential work, and for her support of the project in many other ways. My editor, Jean Thomson Black, has been especially helpful and supportive of this project. I am also grateful to the Institute of Ecology and the Department of Zoology of the University of Georgia for providing me the context and the base from which to teach ecology to generous and often brilliant students. Finally, I would like to acknowledge with appreciation permission to reprint materials granted by Addison-Wesley, the American Institute of Biological Sciences-Bioscience, Cambridge University Press, Chapman and Hall, Clarendon Press, Ecological Society of America, Elsevier Science-NL, Evolution, W. H. Freeman, Harper and Row, Kluwer Academic Publishers, National Aeronautics and Space Administration, Princeton University Press, SPB Academic Publishing, Springer-Verlag, Swedish Natural Science Research Council, University of Chicago Press, University of Georgia Press, Wadsworth Publishing, Peter S. Ashton, Dwight Billings Estate, Helmut Lieth, Gene Likens, Harold Mooney, T. R. E. Southwood, and Richard Westmacott.

Introduction

My objective is to present information on the natural and built environments in a form that will give the general reader an organized way to think about the environment. The information is derived from the environmental sciences, which include geology, physical geography, hydrology, the atmospheric sciences, ecology, anthropology, biology, and human ecology. I will call the organized way to think about the environment *environmental literacy*. In this book environmental literacy is built upon a foundation of scientific concepts.

What is environmental literacy? How can it be developed? What does it mean to take a scientific approach to environmental literacy and what are the limits to a science-based approach? Discussions of these questions will provide an overview of the book.

David Orr, in his book *Ecological Literacy* (1992), begins by stating that literacy is the ability to read. Literacy, like numeracy, the ability to read numbers, is a mark of education. A person who is literate can understand what is written and place it into a context of meaning.

But environmental literacy is more than the ability to read about the environment. It also involves developing a sense of the spirit of place. Orr refers to a sense of wonder. The ancient Chinese made a science of the sense of place, which they called *feng-shui* (Eitel 1984). When my students and I go into the field, I tell them that our first task is to learn to read the landscape. I show them that the landscape is a text that informs us about its capacity to produce and support life, its history, and what organisms are likely to be present. But for me, at least, environmental literacy connotes more than knowing the names of the organisms and understanding geomorphology. I also emphasize feeling the landscape through all the senses. This feeling of place distinguishes each site and makes a place special and memorable.

Environmental literacy begins with experience of the environment. In my case it began in scouting and led to an intense study of camping and woodcraft over about three years. I was fortunate to have the natural landmarks of the Chicago region to explore—the Indiana Dunes and Turkey Run State Parks, the river canyons, the oak-grass savannah woodland, the northern forest of Michigan, and Lake Michigan itself. This region and these landscapes were a birthplace of American ecology (Cowles 1901), although I only learned of that history after I had chosen ecology as a profession. My students tell me similar stories. Scouting, family camps in the national parks, volunteer work as naturalists all figure in these young peoples' backgrounds and provide the direct experience that leads to their study of ecology and environmental science.

Experience is the trigger for environmental literacy. It ignites the curiosity and tests the muscles. It teaches us that we live in a world that is not of human making, that does not play by human rules. We call this world *nature*. To build environmental literacy, it is necessary to go beyond books and libraries and experience nature directly. Only then do we gradually come to recognize a depth and complexity in nature that continually challenge and surprise us. Nature has no purpose, as we humans define purpose, and we soon learn that it does not care about us one way or another. Foolish and risky action in nature can lead to our death. As Paul Shepard (1982) shows, life in nature is one route to human maturity.

From the initial experience with nature the road to literacy runs in a variety of directions. In my case it led to the study of ecological science and ultimately to studies in most of the world's great ecological formations. For others the road leads to a world of imagination, of thought and spiritual growth. The products of such travels are the great paintings of nature, the music, the dance, the books of philosophy and religion, and the life style of the naturalist. You can distinguish a man or woman of the forest, the desert, the mountains, or the sea from the farmer and the city person. Nature marks the human.

Still others have expressed environmental literacy in a life of action. You will find these people in the Congress and other branches of government, in business and industries. Often they are working for change in environmental laws and policies. These people link the health and condition of the environment with the health and conditions of human societies. They return to nature to recharge the spirit; they buttress their arguments with the facts and principles of science, but their life is one of action, or dispute, or the attempt to move what is toward what might be.

Finally, there are the countless professionals who actively work in the environment, rebuilding and constructing nature to fit their vision of an ideal state. Professionals base their work on scientific principles, but the blueprints from which they build are designed for human purposes. Often these purposes are so deeply informed by experience with nature that they fit well into the natural world and it takes an experienced environmentalist to detect the artificial. In other cases, the design reflects custom, tradition, artistic fashion, or economic costs.

All of the people mentioned above would consider themselves environmentally literate individuals concerned about the environment and doing what they can to protect and support the natural world. Environmental literacy can be expressed in many ways. Experience, when combined with scientific study based on imagination, intuition, and disciplined thought, sometimes results in insights that are profoundly different from those of the professional or the activist. In such cases conflict often arises and we must resort to the skills of the conciliator. Where there is no solution, it is best to sit back and allow natural processes and needs to emerge more clearly before the case is reconsidered.

What Does a Scientific Approach to Environmental Literacy Mean?

As I have indicated, I came to ecological science from an initial experience in nature, and I have expressed environmental literacy through scientific research and teaching. But sometimes science can seem anti-environmental, especially when its discoveries are developed with no understanding of social or environmental needs. How does science contribute to environmental literacy?

Science is the study of the material world, the world of matter, trees, animals, rocks, and soil. But it is much more than a listing of the materials one encounters on a walk through a meadow. The scientist searches for patterns of relationship between natural objects and processes. For example, the homeowner in northern Georgia may routinely observe robins, Carolina wrens, and brown thrashers in his or her shrubbery, whereas a sighting of a red-tailed hawk would be noteworthy. Probably the squawking blue jays and mockingbirds would have signaled the hawk's presence before the homeowner looked into the sky. Suburban homes in the southern United States are surrounded by a bird fauna characteristic of ecological "edge," which in the wild forms a border between forests and grasslands. Suburbanization has created an enormous amount of edge habitat and edge-loving species have increased their range as a consequence. The ecological scientist would note the co-occurrence of these bird species and would expect to find several members of the edge bird community wherever one member was encountered. The red-tailed hawk feeds on squirrels, rabbits, rats, and birds, all of which are abundant in suburban areas, but it probably is disturbed so frequently that the edge environment of suburbia is not a suitable habitat. We would not expect to see the red-tailed hawk there, and therefore its presence is cause for comment. This connection of species and habitat is what is meant when we say that we are searching for relationships among the patterns of nature.

The scientist reporting a pattern observed in nature will be cautious. He or she will test the pattern by repeating the observations over and over under different conditions. Scientists know that their observations will be challenged by other scientists. All ideas in science are treated this way. Only those that withstand the challenges deserve to be used to build theory and application. Usually it is not a matter of being correct or incorrect. Rather, we note how the patterns vary and we begin to understand how reliable our observations are. This enables us to say how certain it is that the pattern will appear under given circumstances. Measures of certainty are enormously important. If we are very certain, we are willing to expend large amounts of money or take risks. If we are not very certain, then we will be much more conservative.

The observation of patterns in nature and assessment of their consistency lead the scientist to generalizations. Many kinds of generalization are used in science. For example, a hypothesis is a general statement expressed as a testable question. Given such and such conditions, we might ask, would we expect a particular organism to appear? If the organism did not appear, we would reject the hypothesis that its presence was related to the given conditions. If it did appear, then we would be more confident of having understood the relationship. The test does not prove that the pattern has been explained, but it increases or decreases our certainty.

In this book I will present environmental science in the form of generalizations called *concepts*. *Concept* is a broad term that includes all kinds of general scientific statements, ranging from scientific laws (such as the laws of thermodynamics) to relatively controversial ideas (such as the functional role of biodiversity). Concept is an inclusive word. It really means generalization. It is not one of the steps in the construction of scientific theory. Because concepts are general and often untested, they may contain contradictions. Concepts are especially useful in a primer because they emphasize the normal pattern of environmental relationships and organize an array of observations, interpretations, and generalizations into memorable patterns.

What Are the Limits to a Science-Based Approach?

A primer that introduces the reader to environmental literacy through a series of scientific concepts has the advantage of being able to draw upon the enormous literature of environmental science. There are tens of thousands of papers, books, and reports on the subject, some more than a hundred years old. But the approach also has some drawbacks, and it is essential that I make them clear at the outset.

Since science is concerned with actions in the material world, it does not include perceptions of the environment expressed though the processes of logical thought or the imagination, which belong to the humanities, nor the perceptions of the spirit. This is a serious drawback because so many representations of environmental consciousness are derived from human thought, imagination, belief, and prayer. Most people on the Earth have recognized their close relationship with and dependence on the environment, and they have expressed these relationships in all sorts of cultural forms. A beautiful example of the intersection of religion and science is the experience of Akira Miyawaki, who developed his description of the potential vegetation of the highly altered, humanized Japanese landscape from the remnant natural forests that have been protected around Shinto shrines and Buddhist temples (Miyawaki 1980–1989). In Japanese culture, the place of worship and the natural environment are coupled into a single entity. Whereas in Western culture religious buildings are separate from the natural environment, in Japan spirit and nature are linked and form a whole. These protected forests provided Miyawaki, the vegetation scientist, a base from which to study the natural vegetation of Japan.

Another drawback of my approach is that it does not give priority to environmental action. For many environmentalists, the state of the environment is such a serious matter that it demands our full attention and commitment. They spurn the scholarly detachment of science and the passivity of meditation, demanding action instead. Action requires comparison, examples, cases that can be defended, plans for the best land use, regulations, and policies. Environmental activists want to use the best science available, but their motivation is aggressive and combative, and frequently action requires that we move beyond science. Although this attitude can become tiresome in some cases and is dangerous in others, it often expresses a deep moral commitment and therefore deserves respect. When nature is attacked with violence, the activist is its defender.

My lack of attention to the spiritual side of environment or to environmental activism will disappoint many readers. I respect this response because I share such feelings. I have tried to build links between environmentalism and environmental ethics by including in each chapter a section on the implications of the concepts. Although these sections will be inadequate for some, I hope that they will enable others to move from the science to the other ways we express our feelings about the environment. Philosophers say that we can't move from the *is* to the *ought*. I don't agree. For me it all begins with experience, the first step into the forest, and everything else flows from that.

A Road Map of the Book

The central skeleton of the book is made up concepts of environmental science. As in the spinal column, each concept is a vertebra that acts with the other vertebrae to hold the body upright. The concepts work together to form whole structures that are ordered and logical. This is a primary contribution of the book. In my view, this form of organization is essential in developing environmental literacy.

Recognizing that a large number of concepts potentially could be included in this survey, I have selected a set that are interrelated and can be used to illustrate broad, overreaching ideas central to our concern about the environment. There are twenty-six chapters devoted to an environmental concept or set of allied concepts. But there is also a higher order in the organization of the concepts (fig. I.1). Each cluster into which they are grouped collects those concepts that represent a particular point of view. There are four clusters. We will begin with a basic set of concepts that undergird all the others. Then we will consider concepts that are concerned with the Earth's surface. The unusual perspective here is that these concepts are organized from the Earth as a planet to the smallest unit of surface —that is, from the top down. The third group of concepts are concerned with populations and individual organisms. The final set groups concepts that deal with the interactions of individuals and species. The final chapter concerns human ecology. Taken together, the concepts provide an integrated introduction to environmental literacy.

Readings

Odum, Eugene P. 1989. *Ecology and Our Endangered Life-Support Systems.* Sunderland, Mass., Sinauer Associates.

Pickett, Steward T. A., Jurek Kolasa, and Clive G. Jones. 1994. *Ecological Understanding: The Nature of Theory and the Theory of Nature.* San Diego, Academic Press.

Ecological Concept Clusters

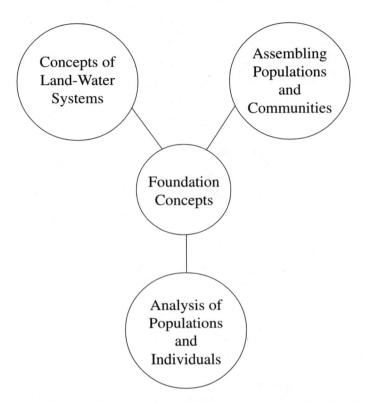

Fig. I.I The four clusters of environmental concepts that organize the book. The foundation cluster is central to the three other clusters.

Cluster I

Foundation Concepts

The first cluster of concepts groups three ideas that collectively provide a foundation for all the others: the environment, ecological system, and ecological hierarchy. Although expressed here as abstract concepts, these ideas are actually quite concrete and find use in many disciplines.

These three "foundation concepts" underlie what many scientists would describe as a "systems approach" to environmental literacy. The term *systems approach* means that I am using language and ideas that relate to a time (the 1950s through the 1970s) when a subdivision of ecology in which I personally have been active thrived. A systems approach seeks to understand nature in the way we observe it, as a whole made up of many interacting parts. It uses systems analytic language and concepts to show how the parts interact in forming whole systems. An alternative to a systems approach would be an interpretation of ecology based on Darwinian evolution. This approach would seek to explain observations of nature through genetics and natural selection.

A systems approach is not a perfect vehicle for our purpose. The main criticism is that it is mechanical and treats nature as a machine. This is a serious objection—nature is above all else not a machine!—but in science I know of no better way to study and discuss wholes than in systems language. Philosophy and religion offer alternative approaches. But in science we face a poverty of tools for exploring the nature of whole systems.

One further comment is in order. We can view nature either as a collection of objects or as process. Examples of objects are forest stands, lakes, and individual organisms. Process involves dynamic flow or change—for instance, the hydrologic cycle, evolutionary change, or the flow of energy. In the second cluster of concepts, concerned with earth systems, I will treat the Earth as an object and its processing of solar energy as process. We can view continua of process over space/time or we can view exchange between objects in space/time. The former emphasizes a field approach to nature; the latter takes a structuralist approach. We will use both approaches to gain an integrated perspective of environmental systems.

1

The Environment

Environment is one of those difficult terms that mean many things to different people. Gerald Young (1986) notes that "environment has become an easy but individualistic code word, utilized in all sorts of contexts, resulting in imperfect, unclear, even misleading dialogues and in the oversimplified reduction of complex matters to a single word." He goes on to discuss the origins of the word. Environment originates in the French *environ* and *environner,* meaning "around," "round-about" and "to surround," "to encompass." These in turn are derived from Old French *virer* and *viron*, meaning "to circle," "to turn about," "around," "the country around," or "circuit." From this etymology, we conclude that environment means the things or events that surround something else.

This thought image implies a passive or neutral object that is surrounded or encircled by its environment. More usefully, we may add to the definition active interaction between object and environment. If the object is an individual organism, then the environmental factors that surround and interact with it may include predators or foods, chemical elements, the soil, or gases in the atmosphere. Figure 1.1 describes this interaction system in a circular diagram, which I have interpreted to Asian audiences as an environmental mandala, after the Buddhist and Hindu geometrical symbol of the cosmos.

The diagram represents the environment of a single plant, but it can be made general by replacing the plant with an animal, a biotic community, or any other complex of living organisms. The environmental factors interact with the living organisms directly, as indicated by the solid arrows, and indirectly through other environmental factors, indicated by the dotted arrows. For example, soil directly provides plants with nutrients required for growth, but it also indirectly influences water availability through soil

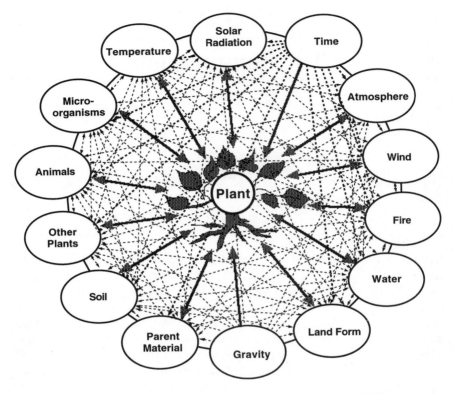

Fig. 1.1 The holocoenotic diagram redrawn from that of Dwight Billings (1952), showing the environmental relationships of a single plant. Solid lines show factor-plant relationships, dashed lines show relationships between factors and, indirectly, the plant. Arrows indicate the direction of the effects.

porosity and water storage capacity, which in turn influence the amount of water required by the plant. Indirect influences may be multidimensional and many steps removed from a direct influence. In addition, the diagram shows that the organism acts upon the environment; the reaction system is multidirectional.

If our object of interest is an environmental system made up of the interacting organisms and the physical factors and chemicals in air, water, and soil, then the environment consists of the neighboring systems that share the surface with it and the broader features that are associated with Earth movement—tide and waves and weather. Hence the environmental system containing a local environment also has a larger environment. The key point in the definition is the encircling or surrounding nature of environment. The system of interest will vary depending upon our purpose.

There is one further common usage that can be confusing. Frequently the natural environment is contrasted with the built environment. What do we mean by these terms?

The Natural Environment

In the phrase "natural environment" we encounter a form of the familiar word, nature. But *nature* is an ambiguous term that is used in a variety of ways and frequently confuses rather than clarifies a discussion. *Webster's Dictionary* (1959) states that nature comes from the Latin *natura* meaning birth, origin, natural constitution, or quality of a thing. The dictionary defines nature as "the essential character of a thing; quality or qualities that make something what it is." By this definition, the natural environment is the environment before human presence or before human disturbance or management. But according to modern ecology, such a definition contains two errors. First, it suggests that the natural world is static and fixed and that humans have disturbed and changed it in some way. Our current understanding of nature as dynamic and changing is quite different, and our definition needs to reflect this.

Second, the definition implies that there was a time before humans when the forest was virgin or the wilderness was pristine. Then humans appeared and changed nature into something else. This outdated model usually places indigenous peoples within nature and then focuses on the change induced by modern or European peoples. Although ecologists accept the idea that natural communities of plants and animals were present before human evolution, such communities demonstrate the widespread impacts of indigenous humans once humans became part of the landscape. Humans manipulated plants and animals by feeding on them, they manipulated fire and the constitution of forests, and they appear to have caused the extinction of fragile large animals. Human impact is not a recent event, although today it is much wider-spread and has more serious repercussions than ever before.

Given these considerations, we can use the term *natural environment* in a practical way to refer to the environment that is not managed by humans for human purposes. The natural environment develops through the interactions of organisms among themselves and the interactions of organisms with the soil, water, and atmosphere. It includes the seacoast and the marshes, the lands that have been abandoned by farmers and which are becoming revegetated, the clumps of forest or grassland that remain from the past, the deserts that no longer support grazing, the landscapes of national parks, rivers, lakes, and the oceans. Natural environment contrasts with environments dominated by human purpose, such as built environments.

The Built Environment

The natural environment is the home in which humans evolved. Human history can be interpreted as the development of tools and skills that manipulated aspects of the natural environment and enabled humans better to survive. The discovery of the stone arrow and spear point, the scraper and fire, must have immeasurably improved our ancestors' quality of life. But even with these inventions, human populations mainly were controlled by the natural environment. And humans, as I noted above, had a reciprocal influence on the environment.

When humans discovered agriculture, the ecological situation changed radically. Agriculture is the manipulation of plants or animals to focus production on humans. Such manipulation is not simple. It probably required centuries of trial and error before humans could reap predictable crops in the periodically flooded riverbeds or on the edges of lakes and marshes where water and grainlike species of grasses and herbs abounded. Agriculture, however, involves much more than production of food. For example, it involves a sedentary life style. The farmer stops moving about following natural productivity and settles down to manage and harvest crops. In doing so, the original farmer created a new form of environment, the built environment. The built environment is made by humans for human purpose and represents patterns appropriate to human culture. Culture is the shared set of beliefs and attitudes of a group of humans that structures how they decide what is right and wrong and how they act. Culture, like environment, is larger than the individual human. The individual is embedded in a culture and may be unaware of the cultural constraints on the way he or she thinks and acts.

The human-built environment of fields and houses had other effects on human populations and the natural environment. For example, agriculture involves management of a surplus. Mobile populations have a limited capacity to store surplus production. They can't carry the surplus very far, and stored or hidden fruits and seeds attract predators and scavengers. In contrast, farmers must store production to carry them through lean periods and to provide the seed or stock for the next growing season. Crop storage requires special containers and this need stimulated production of ceramics and basketry crafts that require skills different from farming. As a consequence, a division of labor among humans began to develop.

Agriculture also involves ownership of land. Mobile people may range over a region that they identify as theirs, but there is little point in defending a surplus that you cannot use. Mobile people roam widely and then meet to exchange gifts and products, select wives and husbands, and create social bonds. Farmers work a specific area of land and use a specified amount of water to irrigate it. The land and its products are related so that the owner of the land receives the value of the product, no matter who tills the soil. The ego interest of the individual, which was more tightly controlled in the mobile society, now has room to grow. Individualism has a purpose. Ownership can be passed to another generation and therefore parentage becomes important.

As agriculture expanded, large sedentary human populations—those who kept the calendar and could tell when farming practices should be carried out in anticipation of the yearly cycle—could evolve. Armed warriors were needed to defend the storehouses and the lands against predatory humans who preferred stealing food to growing it. These activities required structures and the first settlements appeared, grew ever larger, and eventually became citylike. The process that led to the modern condition was under way thousands of years in the past. Our life style is the extension of these ancient trends.

The point of this discussion is not to recall the history of agriculture and settlement but to point out that the farm, pasture, village, and city are all environments. They differ fundamentally, however, from the environments in which humans hunted and gathered food hundreds of thousands of years ago. These new environments were created by humans for specific purposes. Although they evolved over centuries and were shaped by the regional patterns of climate and soil, they are purposeful in a way that natural environments are not. Nature is purposeless, in a trajectory of continuous change, beginning not long after the Earth came into being 4.7 billion years ago and ending when our star, the sun, stops producing energy. Life has evolved and will continue to evolve to fit the existing environmental conditions. In contrast, the environments humans construct serve specific purposes. Human imagination, will power, and purpose are expressed though built environments (Rapoport 1977).

In some ways, however, the built environment functions similarly to the natural environment. It forms a reaction system in which humans play the central role that the plant played in figure 1.1. The human construction of an environment has several consequences. First, the built environment contains, as one element, the human purpose that motivated its construction. It has been constructed for a purpose and the design used to guide its construction was selected from all possible designs as a compromise among competing needs. The humans who use and live within these built environments incorporate the purposes of their construction into their being, and the humans become shaped by the built environment. As an example, visualize the street plan of American cities in the Middle West. The streets form grids regardless of topography. The rectangular grid is ideal for automobile traffic but not for other aspects of urban life. It makes formation of neighborhoods more difficult; it is dangerous for children; it brings noise and gaseous pollution into residential areas. Affluent citizens react to the negative features of these grids by leaving them for suburbia, where they construct rolling, curved roads with parks and pedestrian ways. Central grids are left to the poor or the urban pioneer who reconstructs the derelict structures of the central city. American city development, as described by John Reps (1965), involves interaction and feedback representing the kind of environmental interactions discussed earlier.

The problem is clear: Human design has a purpose, but it is expressed in an environment that consists of multiple interactions and feedbacks. Any single purpose generates multiple responses which alter that purpose. If the result of human design is a built environment, this environment has its own complex of interactions, most of which are unknown and unpredicted. As a consequence, the environment requires continuous reconstruction. This is a natural pattern of response that would be characteristic of any organism. What makes it different where the human is concerned is its scale. Humans often try to compensate for the problem of the construction being embedded in other environments by increasing the size of the construction so that it incorporates the natural environment. The human urge to conquer the Earth is a manifestation of this attitude. If humans somehow could control everything, many people think, the problems would be reduced. This belief is based on a misunderstanding of environment. Indeed, by eliminating processes that have stabilized the Earth systems over millions of years, we can be certain that the human-built environments will be less and less able to sustain human life.

Implications

The environmental concept implies a merger of self and other. "Other" means the other members of one's group, other organisms and the other nonbiological forces that potentially affect the self. Consider the human body, in which the cells of the skin, blood, intestinal tract, and other organ systems are replaced on a regular schedule. For instance, the surface cells of the mucosa of the gut wall turn over once every four days. The new cells that replace those that have been sloughed off and discarded are constructed from matter and energy obtained from the organism's environment. This means that the body is physically linked to the Earth through the constant reconstruction of its tissues. The molecules that become "me" come from the food I eat, the water I drink, and the air I breathe. And I affect the environment through the waste products I produce from my digestive tract and kidneys and the carbon dioxide I exhale from my lungs. There is no divide in existence. The self and environment form a whole.

By stating the environmental concept in this way, we are describing a paradox: a unique and free individual that is completely coupled to an environment. The environment consists of air, water, energy, matter, other organisms, other humans, the built environment, and culture. Although individuals are both unique and free to imagine and act, at the same time we are tightly embedded in an environment and, as humans, a culture.

We can't resolve this paradox by arbitrarily deciding that one or the other side of the contrast is untrue, unimportant, or irrelevant. Rather, we might resolve it by expressing our independence in appropriate ways within an environment and a culture. This shifts the question from "How do I resolve the paradox?" to "What are appropriate actions to take within a given environment?" Obviously, the actions one takes to live in a forest are different from those taken in a desert. "Appropriate" actions are those that would preserve and enhance both the environmental factors one interacts with, within one's private interaction diagram, and those associated with our individual survival and well-being. It is self-interest to be concerned about one's environment. If we are coupled to our environment, then it is foolish to ignore it or, worse, to trash it. The impact back on us is direct. There is no escape from ourselves.

This resolution of the environmental paradox gives us reason to be optimistic. We have the capacity to rebuild environments so that life and the Earth and water have vibrant relationships. We can even rebuild environments to enhance biological diversity. We can make the environment more beautiful. We can manage ourselves so that human life is productive and meaningful and most individuals are able to develop self-respect. We can eliminate crime, hate, and paranoia. These harms are not given elements of being human. They are a matter of will.

Readings

Bartuska, Tom J., and Gerald Young, eds. 1994. *The Built Environment.* Menlo Park, Calif., Crisp Publications.

Bates, Marston. 1967. "A Naturalist at Large: Environment." *Natural History* 76, no. 6: 8–16.

Brandon, Robert N. 1992. "Environment," pp. 81–86. In E. F. Keller and E. A. Lloyd, eds., *Keywords in Evolutionary Biology.* Cambridge, Mass., Harvard University Press.

Dubos, Réné. 1973. "Environment," pp. 120–27. In P. P. Wiener, ed., *Dictionary of the History of Ideas: Studies of Selected Pivotal Ideas.* New York, Scribner's. Vol. 2.

Howard, Philip. 1979. "Environment," pp. 82–85. In *Weasel Words.* New York, Oxford University Press.

Young, Gerald L. 1986. "Environment: Term and Concept in the Social Sciences." *Social Science Age* 25: 83–124.

2

The System

The preceding chapter identified a complex of organisms and environment that interact to form a whole. In this chapter I will refer to this whole as a system. A system is an object that is made up of subsystems or components which interact in such a way that they have, collectively, a recognizable wholeness. I want to emphasize that we are concerned with wholes which can be analyzed into component parts. Science studies individual organisms, flocks and herds of animals, lakes and patches of meadow, the boreal forest, and oceans. Each can be considered as a whole, and we will do so here.

Definition

The word *system* is used in many different contexts. The dictionary defines it as a collection of objects organized according to a plan and forming a unity or a whole. This definition has several parts. First, there is a collection of objects in a system. Although these objects can be separately identified, they also interact with each other. Second, these objects or processes are organized. Organization implies a connection between the objects or processes. Third, there is a plan of organization. That is, the connections between subsystems are repeated in time and space. Finally, the objects form a whole. This means that the objects or subsystems that make up the system have properties that pertain to themselves. They also have properties that make them an integrated whole. It is important to note that the properties of the whole are not the mere sum of the properties of the subsystems, as the above sentence may suggest. The properties of the whole cannot be predicted from knowledge of the parts. They emerge from the whole itself.

As this is a difficult concept to understand, it may be helpful to give an example. The classical illustration of the part-whole concept is the water molecule. Water is made up of atoms of hydrogen and oxygen. Each chemical element has specific properties that are unique to it. Oxygen when combined with iron, or hydrogen when combined with nitrogen, acts differently than when oxygen and hydrogen are combined. It is unlikely that we would predict the properties of water from knowledge of the elements separately. What we know as the familiar properties of water emerge from the combination of these two elements.

Similarly, a forest is made up of many species of organisms, each having special properties. We cannot predict the properties of the whole forest from knowledge of these individual species. A forest has unique characteristics that emerge from the interaction of the species with each other and with their physical environment. For example, the productivity of the forest is balanced by its metabolism. This balance is not the sum of the balance between the growth and metabolism of each species or each individual, because they interact with each other and change each other's individual performance. Interaction produces something new and different that is not seen in the parts individually.

I have suggested that a system is an object in space-time. In common English, an object is defined as a separate entity with visible boundaries. But this commonsense definition is the result of a visual focus. It conflicts with our understanding of objects as environmental complexes. From now on in this book, when I use the words *object* or *system*, I will be referring to an environmental whole consisting of the living part and its environment.

This decision does not, however, eliminate the problem of locating the system boundaries in space-time. Environmental systems usually do not have well-defined boundaries. Rather, they blend into the neighboring systems so that their characteristics change over a gradient representing the zone of interaction of a system with its neighbors. We often say that the system boundaries are fuzzy. Ecologists have given this zone of interaction a specific name, the ecotone, because it is so common and so important. Many organisms are characteristic of ecotonal habitats. In Chapter 1, I noted the robin, brown thrasher, Carolina wren, and blue jay as bird species found in edge habitat around suburban homes in eastern North America. With our understanding of the systems concept, we can now refer to these species as characterizing the ecotone between forest and shrubland or forest and meadow systems.

Based on this analysis, an environmental system can be defined as a weakly bounded object made up of living and environmental components and processes that interact to form a whole. The environmental system is open because it receives inputs from and gives outputs to other systems that make up its environment. In 1935 Sir Arthur Tansley, one of the premier ecologists in England, coined the term *ecosystem* for this kind of system. Tansley emphasized that ecosystems are physical systems and that they "are the basic units of nature on the face of the earth" (Golley 1994). Tansley's term caught on among ecologists and environmentalists, and in a 1989 mail survey the members of the British Ecological Society ranked the ecosystem as the most important concept in the science of ecology.

Ecosystem, like environment, has both a general and a specific meaning. It can refer to a particular kind of system on a site and it can refer to systems in general. For example, we could speak about the Lake Michigan ecosystem specifically or we could refer to lake ecosystems in general. When the word *system* or *ecosystem* is used, it is important to make clear the reference of the term. In this book I will use *ecosystem, ecological system,* and *environmental system* as synonyms.

An Example

An example of a system is a small lake or pond that forms in a topographic low area where water can accumulate (fig. 2.1). The size of the lake and the volume of water depend upon the balance between inflow from rain, snow, streams, or springs and the loss of water by evaporation into the atmosphere and flow from the lake either above or below ground. The study of water balance in a lake must take all these processes into account. The lake water and the ground beneath it have a chemical and physical structure that influences the nature of the lake. If the lake is rich in nutrients, then it may be highly productive and even clogged with growing plants. If it has few nutrients, it may be clear and deep blue in color. The lake also contains living organisms that interact with each other, the water, and the sediments. The organisms range from fish to microscopic plants, animals, and microorganisms such as bacteria. Although we may choose to examine one single element in the lake ecosystem—the fish, for instance—the character of the fish fauna is a result of many interactions of the processes and components within the lake ecosystem, as well as the environment of which the lake is a part. When we say "lake" we mean all of these components and processes acting together as a whole. Most of us have little trouble deciding where the lake begins and ends or measuring the inputs and outputs across the margin. For the scientist studying lake systems, however, this boundary is dynamic, moving up and down as the water level rises and falls. Many chemical and biological processes take place at the margin of the lake, and therefore it is an important ecotone.

The concept of ecosystem can be applied to built environment systems in the same way we applied it to the natural environmental system of the

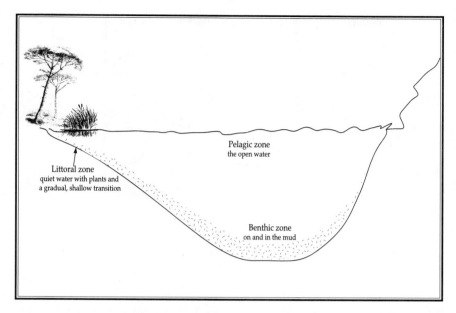

Fig. 2.1   Cross-section of a pond showing various zones of the pond ecosystem.

lake. A particularly interesting example, which is actually a system built on ecological principles, is Biosphere 2, located in the Arizona desert near Tucson (Alling and Nelson 1993). This imaginative experiment was publicized as a step toward building a space colony on Mars. It involved the construction of a sealed glasshouse in which a variety of systems such as ocean, marsh, rain forest, grassland, and desert interacted with cultivated fields and complex industrial systems. In the first experiment, eight biospherians occupied the three-acre complex with selected plant and animal species and a collection of cryptic and unsuspected fellow inhabitants for two years. Alling and Nelson comment that it took almost the entire two years for the humans to fully adapt to the biosphere, but eventually it felt like home. A review by a scientific committee in 1996 identified a series of problems that emerged during the first experiment: the atmospheric oxygen concentration fell from 21 percent to 14 percent, N_2O concentration in the air rose to dangerous levels, nineteen of twenty-five of vertebrates went extinct, all pollinators went extinct, the majority of insects went extinct, and water systems became loaded with nutrients. The conclusion of the committee was that "at present there is no demonstrated alternative to maintaining the viability of the Earth. . . . Earth remains the only known home that can sustain life" (Cohen and Tilman 1996).

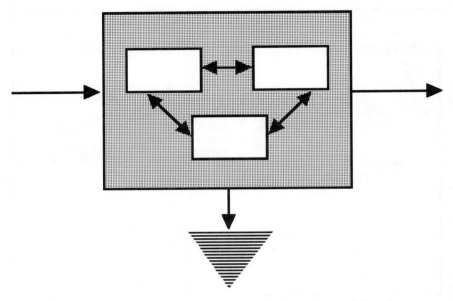

Fig. 2.2 A diagram of a simple system, showing inputs and outputs to and from the system, the output of energy as heat below the system box and some components within the system that are linked through the exchange of energy, matter and information.

Language

I will use a language and symbols derived from Forrester (1971) and H. T. Odum (1983) to discuss environmental systems. In this symbology the system is identified as a rectangular box (fig. 2.2). Its limits are defined by the straight lines that define the boundary of the system. Inside the box are the components or subsystems relevant to the process being studied, symbolized by rectangular boxes of smaller size. The connections and influences between components are shown by arrows, with the direction of the arrow showing the direction of the influence. Arrows pointing in two directions mean that the subsystems interact through feedback processes. Arrows also enter and leave the system and these arrows indicate coupling of the system to other systems in its environment, which are not shown in the figure. Finally, the symbol beneath the rectangle indicates the dissipation of energy, matter, and disorder from the system.

The systems-analytic framework is not only conceptually useful but also permits us to organize and manipulate information about systems on an analog or digital computer. In the early days of systems analysis, ecologists constructed analogs of systems where capacitors or rheostats took the place of components and controls and the flow of electrical energy was an analog of flows of energy or materials through the ecological system. These analog computers were limited in the number of components and interactions they could simulate, and they were quickly replaced by digital computers that can be programmed to handle extremely complicated systems, such as the global climate or the world economic system.

The systems approach is useful because it is abstract and general. It allows us to discuss lakes, forests, and organisms in the same language. The danger inherent in the systems approach is that we may use false analogy, moving from one system to another inappropriately. In extending knowledge across systems, we need to demonstrate their similarity or define their differences.

The opposite of systems thinking is to treat each ecological whole as unique and independent. Natural history sometimes treats organisms in this way, especially when biologists are exploring the role of evolution and natural selection. The reason they do this is that natural selection acts on the individual organism, so the organism and its population are the key systems in evolutionary ecology. However, an individual organism also can be treated as an ecosystem when the interaction of the biotic element and its environment is of paramount interest.

Implications

The choice of systems language to discuss the science of ecology has implications for how we think about the natural world. Using the word *system* might lead us to think that nature is like a machine and that we can approach it from a machine point of view. Living in the modern world where we have almost no contact with nature makes machine thinking inevitable and normal. But that is *not* how I am using the word here! By system, I mean an organized whole that can be subdivided into components. In other words, I am assuming that nature is organized order.

In order to understand the implications of the system concept, it is necessary to examine two different topics. The first concerns the way scientists think about whole and parts. The second concerns other than scientific ways of thinking about whole and parts.

In science the initial period of exploration often yields patterns of events in nature that are consistent across space and/or time. A consistent difference between the amount of surface litter lying on the soil of two forest stands is an example of an observed pattern. Let us say that in one case the amount of litter or dead leaves and twigs is about 500 grams per square meter, while in the second forest the amount is 1,500 grams per square meter. These are substantial differences and they elicit the question, Why do these stands differ in the amount of litter per area? The scientific way

to address this question is to identify the subsystems that make up the litter and examine those subsystems of significance to the question. For example, we might measure the abundance of earthworms in each forest because earthworms participate in litter decomposition. We might gather
small amounts of litter in nylon bags and place them on the ground. By harvesting these bags at intervals, we can weigh them and determine the
amount of leaf and twig tissue remaining after decay has taken place. There
may be a number of similar studies that we would undertake to answer the
question. Some of them, such as the breakdown rate of litter, might require
us to investigate a process at an even finer scale. We might want to examine
the rate at which fungi attack the different kinds of organic matter in the litter. And so on.

In this example, we have gone deeper and deeper into understanding
the process of litter breakdown in our search for an explanation of the difference between forests. This process is called *analysis* and it is the primary perspective of scientific work. Logically, analysis involves posing a
question, identifying the parts of the system, examining the parts in the
context of the question, deciding how effective the analysis is in answering
the question, and usually further subdividing a part and repeating the
process in deeper detail. At each stage new questions emerge and the research can go on for a long time.

A less common approach in science is to consider the question at successively broader scales. In the above example, we could have asked if the
observed difference in the amount of litter influenced the rate at which
forests developed in the region. We might note that the forest with less litter was younger than the forest with more litter. We might hypothesize that
the difference reflects the maturity of the stands and that ultimately the
large quantity of nutrients stored in the litter would allow other species to
enter the stand and the forest would become richer in species. This approach is called *synthesis*. It has the goal of putting components together
into wholes and then drawing conclusions about the results of synthesis.

It is obvious that both analysis and synthesis are necessary for a full understanding of phenomena. In the United States, however, the emphasis is
on problem solving, and therefore the analytical approach is most commonly used. In our culture the context is either assumed (and thus of little
interest) or is considered outside the area of responsibility of the problem
solver. Because of this bias, the methods of analysis are exceptionally well
developed in all the sciences. The only synthetic method developed in science has been systems dynamics. It did not develop rapidly until the ad-
vent of the computer; now it is a widely used methodology.

Analysis and synthesis are also employed in other approaches to understanding the environment. However, the creative insight that leads to a

unique pattern expressed in language, painting, or music, for example, tends to come out of the synthetic experience. The artist examines, thinks about, and imagines the scene, and then the image appears in the mind. By means of technical skill, the artist then transforms this mental image into the image that you and I appreciate. Our understanding of the environment likewise can be transformed by the experience of viewing the image created by the artist. The emergence of the whole—the painting—through the mind and skill of the artist and its influence on the viewer represent a sequence of synthetic processes. Of course, our appreciation is influenced by our culture and training, which lead to certain expectations as we view the painting, just as the artist's training and sensitivity influence the ability to create the painting in the first place. The painting forms a point of interaction between viewer and artist, a context that both carry with them.

Because the holistic process is so fundamental to creativity in the fine arts and humanities, they are a crucial part of environmental studies. Environmental science can be merged with these other forms of scholarship to give the full range of understanding of the environment. But here our focus is on science, and we can do little more than to tip our hat to these other topics.

Readings

Becht, G. 1974. "Systems Theory: The Key to Holism and Reductionism." *Bioscience* 24: 569–79.

Golley, Frank B. 1994. *A History of the Ecosystem Concept: More Than the Sum of the Parts*. New Haven, Yale University Press.

Lazlo, Ervin. 1972. *Introduction to Systems Philosophy: Toward a New Paradigm of Contemporary Thought*. New York, Gordon and Breach.

Miller, J. G. 1984. "The Earth as a System." In *Models of Reality: Shaping Thought and Action*. Mt. Airy, Md., Lormond.

Nelson, Mark 1993. "Using a Closed Ecological System to Study the Earth's Biosphere: Initial Results from Biosphere 2." *Bioscience* 43: 225–36.

Odum, Eugene P. 1989. "The Ecosystem," pp. 38–66. In *Ecology and Our Endangered Life-Support Systems*. Sunderland, Mass., Sinauer Associates.

Patten, Bernard C., and Sven Jorgensen. 1995. *Complex Ecology: The Part-Whole Relation in Ecosystems*. Englewood Cliffs, N.J., Prentice Hall.

Shuggart, H., and R. O'Neill. 1980. *Systems Ecology*. Stroudsburg, Pa., Dowden, Hutchinson, and Ross.

3

Hierarchical Organization

In Chapter 2 I developed the concept of an ecological system, characterized by specific organisms and abiotic features, which blended into other systems surrounding it. Taking a geographical perspective, we could view a land surface as covered by several of these environmental systems and their ecotones. The surface of the Earth is made up of a mosaic of ecosystems of varying kinds. We only need to look out of the window of a low-flying airplane to verify that this description accurately represents the visible patterns on the ground.

In this chapter I will develop a vertical perspective of environmental systems, in contrast to the horizontal view discussed in the last chapter. If we were traveling by spacecraft rather than airplane and our trajectory was toward deep space, our view of the land surface would change as we moved farther and farther from Earth. First, the mosaic of environmental systems we observed from the airplane would merge together and the details would disappear. We would only be able to see large-scale patterns of forest, grassland, and water. Houses, roads, and small fields would be invisible. Eventually the Earth would become a ball in space, and we could only recognize continents, ice caps, and oceans. Finally, as we returned to Earth, details would reemerge until close to home we might recognize our town, our street, and even our house before we landed. This thought experiment demonstrates that the identification of environmental systems depends upon the scale of perspective. At the finest scale we could be viewing the familiar systems we live among; as the scale becomes more coarse, our view gradually may shift to the whole Earth.

These perspectives give us two ways to examine nature: we can take a horizontal perspective, viewing the surface as a mosaic of environmental systems, or a vertical one, which collects or subdivides the environmental

18

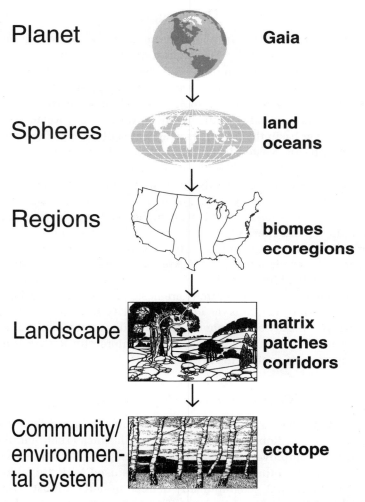

Fig. 3.1 An illustration of a hierarchical perspective showing five nested divisions of the Earth ecosystem, or Gaia. The terms *planet, spheres, regions,* and so on indicate the divisions of the hierarchy. *Gaia, land and oceans, biome, ecoregion,* and so on are used for these divisions in environmental science.

systems into supersystems or subsystems, depending upon the questions we are pursuing. The latter process organizes environmental systems into nested sets (fig. 3.1). If we begin with the largest system we know, the planet Earth, and apply the process of analysis, we can recognize continent-sized areas of ocean and land, which in turn are made up of landscapes, which in turn are composed of the local environmental systems that we encounter daily.

Definition

The scheme of organization in which systems are nested within one another is known as a *hierarchy*. The highest-order system in the hierarchy contains subsystems (as shown in fig. 2.2), each of which contains subsubsystems, and so on. The criteria we employ in distinguishing members of the hierarchy can vary, depending upon the question being asked. If, for example, we were concerned with tracing the flow of surface water, we might begin with a river basin that occupies a large area of land, such as the Mississippi River basin. Nested in the Mississippi River basin are river systems, such as the Missouri River, the Ohio River, and the Tennessee River. Each of these rivers contains smaller basins. Eventually, we might examine the smallest headwater watersheds, of which there are thousands. Each basin or watershed is nested in the next-higher system. The only requirement in building this hierarchy is that we use the same criteria—the flow of water in surface streams—to identify the systems at each scale.

Hierarchy theory is a tool for thinking about nature. The primary requirement for using hierarchy theory in the context of environmental systems is that the levels of the hierarchy are nested one inside the other. Each level of the hierarchy contains the lower levels within it. The nestedness of the systems is a purely scalar phenomenon. It does not mean that higher units control the lower units, as in a military hierarchy. Rather, it means that the higher units contain the lower units.

Criteria for Establishing Hierarchies

One of the key questions in hierarchy theory is how to aggregate or decompose the nested set. Robert O'Neill and his colleagues (1986) emphasize that the criteria employed in aggregation or disaggregation must be self-similar—that is, that the same criteria must be used to distinguish each level of the hierarchy. A geographically defined nested set must always be referenced to an area of land or water. This is what "geographical definition" means. One cannot shift from geographical to biological criteria in constructing a hierarchy. For example, the small-lake ecosystem discussed in Chapter 2 can be decomposed into an open-water subecosystem and a sediment subecosystem. Both contain living organisms interacting with a chemical and physical environment, in the same way as the lake did as a whole. The open-water subsystem might be further decomposed into near-shore, deep-water, and surface-water subsubecosystems. All of these systems and subsystems preserve the major ecosystem property of an interacting biota and physical-chemical environment. Similarly, the lake is part of a watershed ecosystem where the lake interacts with rivers and terrestrial systems of fields and forest. It would not be legitimate to include a level in this hierarchy that would focus on the behavior of fish because this category is not self-similar to the watershed, lake, open-water nested set.

The principle of self-similarity is important because environmentalists commonly disaggregate a lake and other systems into biotic and geographic subsystems and mix the two. It would be acceptable to make such a division if one were not interested in the hierarchical relationships of the lake. And it would be perfectly acceptable to decompose a hierarchy on biological criteria, with the water and sediment treated as the biota's environment (MacMahon et al. 1978). The biota could then be divided into plankton, benthos, and fish or into groups such as herbivores, carnivores, and decomposers. This form of decomposition emphasizes the biological rather than the geographical features of the system, which might serve another purpose of the investigator. But the geographical and biological are two different analyses, and they cannot be mixed logically to create a hierarchical structure.

Advantages of Hierarchical Organization

The horizontal coupling between subsystems often is relatively loose, being limited to the inputs and outputs of each system to the next. This means that each system can express its own behavior and be semi-independent, as well as being bound into a system of relationships. Vertical couplings also may be loose, and system independence is possible in that direction as well. Even though a popular ecology rule says that everything is connected to everything else, actually loose connectivity means that there are many discontinuities in the hierarchy (Allen and Starr 1982). Connections between successive levels do not necessarily mean connections across systems within a level. Ecologists find that connectivity in real-world systems is usually less than 20 percent. Discovery of discontinuity is an important step in studying hierarchical systems.

The property of independence in hierarchical organization increases the stability of a system. This advantage is illustrated by the watchmaker story (see Pattee 1973, p. 7). Two watchmakers are assembling fine watches with ten thousand parts. Each is constantly interrupted to answer the phone. The first has organized his work hierarchically, with 100 subassemblies of 100 items. The other has not organized the work in this way but instead proceeds linearly, attaching part 1 to part 2 and so on. Each phone interruption causes the system to fall apart since the watchmaker cannot recall where he was in the process. If the interruptions come at about 150 unit-assembly intervals, the first watchmaker eventually will assemble a watch.

The other watchmaker will always fail. Hierarchical organization makes the assembly process more stable and more efficient.

The property of independence also permits evolution and change to occur within the subsystems made up of individual organisms and species. One subsystem can evolve and change its method of processing information, for example, but it cannot change the inputs or outputs coupling it to other subsystems. After a certain amount of change, the subsystem may become more efficient and more competitive, and then could change its relationships with other subsystems. These changes might appear as jumps to us, but actually they would be the result of the accumulation of many small changes within the subsystem over time.

The Approach

In this book we will visualize the Earth as a spatial hierarchical system, with the highest level being the planet. We will focus on the energy dynamics of the Earth and then examine its four spheres: the lithosphere, atmosphere, hydrosphere, and biosphere. The Earth system will then be decomposed into a series of finer-scale subsystems, which we will call biomes, landscapes, watersheds, and ecotopes. (These terms will be defined in separate chapters below.) This geographical hierarchy will form the basic organizational structure of the second cluster of the book, and I will present appropriate environmental concepts within this spatial hierarchy.

Is there an advantage to progressing through the hierarchy from the top down as opposed to moving from the bottom up? I have chosen a top-down approach because it follows the method of analysis by which scientists dissect a problem. We identify the problem, consider the components or parts of the problem, and then come to a conclusion about how the interaction of the parts explains the problem. To move from the top down does not require advanced understanding. It only requires that we identify the components and their interactions. The alternative bottom-up approach requires that we understand the plan or design of the system, or that we have a theory which shows us how we are to put together the components to create a higher-order system in the hierarchy. That is, we require *assembly rules* to construct the system from its parts. This is a useful direction to take in some areas of science, and I will employ a bottom-up approach toward the end of this book when we deal with the construction of populations and communities from individuals. A bottom-up approach is widely used in ecology and is especially helpful when focusing on only a few levels in a hierarchy.

Implications

Although the spatial hierarchy is a powerful tool for organizing information about ecological systems, it presents a problem that we must recognize explicitly. The term *hierarchy* is widely used in describing human social organization. In this sense, it is based on control and the distribution of power. Humans in higher levels of a hierarchy such as a military unit control those in the lower levels. As a consequence, hierarchy theory is objectionable to people who are fighting prejudice and bias. Feminists, for example, tend to bristle when they hear the term because they know that human hierarchies usually have been used to defend male dominance. We can understand these feminist arguments, but it is difficult to think of another way to organize a subject as broad as environmental science.

Carole Crumley (1987), an anthropologist at the University of North Carolina, has used the term *heterarchy* in place of hierarchy. She defines heterarchy as a ranking of system elements in different ways. Thus, a heterarchical individual may be a member of many hierarchies. Heterarchy may be an improvement over hierarchy. Certainly in my usage of hierarchy there is no implication of rank, where those systems higher (or to the left or right) in the hierarchy have more or less significance than those at another level. The nested-set hierarchy only organizes or classifies systems in a scale.

The recurring idea that emerges from our three foundational concepts is relationship or connectedness. No process or object is isolated. All are linked and, even more important, are embedded within other processes and objects. Environment emphasizes the connections between organisms and other organisms and the physical chemical elements of the environment. System emphasizes the connections between components of the system and their exchange of materials and energy to form a whole. Hierarchy emphasizes the embedding of one whole in another. The overall pattern is a complex relationship from which there is no escape. Yet these relationships are not a denial of creative freedom in nature. Relationship is not necessarily control. Rather, it tends to channel and shape creativity so that wholes evolve and change over time.

The centrality of relationship in the natural world contrasts markedly with the concept of individuality and independence in modern culture. This contrast marks one of the key differences between what I call the "environmental" and the "modern" approach to life. The environmental approach places connection at the heart of all decisions and seeks improvement in the human condition within nature so that both the natural systems and human well-being are sustained. This means that human growth and demand for more material objects are limited by nature's capacity to sustain them.

The modern approach emphasizes the individual human and is concerned with finding the resources adequate to meet the demands of individuals. These demands may be expressed in hierarchies of power where those in the most powerful position are most likely to meet their expectations. Individual demands are generated by human creativity, so that they are endless and in one sense incapable of being fulfilled. The frustration of this situation is defined as a good, because it is considered a driver of economic and social progress.

Nature provides resources and services for the individual. Since ecological connectedness is a real property of nature, unlimited demand eventually results in the breakdown of natural processes. Destruction of nature spurs a search for substitute materials or processes. Invention of substitutes is also defined as good and a mark of progress. Thus, pursuit of the positive features of the individualized society causes greater and greater ecological destruction.

Environmental ethics derives from the conflict between these points of view. In my opinion, it was during the period 1940–1950 that the individualized point of view came to dominate the relational one. Before 1940 there were viable human sustainable systems in natural landscapes all over the earth. After 1950 these systems were assaulted by the dominant societies and destroyed almost everywhere. Today even the remnant natural landscapes are being destroyed to provide resources for a human population that is growing out of control.

Environmental ethics teaches that environmental destruction is wrong and points toward relationship and sustainability. Human creativity can be focused on enhancing the relationship between the human and natural worlds and forestalling the impending tragedy. Our task here is to understand how ecological systems operate so that this intention can become practice.

Readings

Bertalanffy, Ludwig von. 1952. "Levels of Organization," pp. 23–54. In *Problems of Life*. New York, Wiley.

Egler, Frank E. 1970. "The Nine Levels of Integration," pp. 122–42. In *The Way of Science: A Philosophy of Ecology for the Layman*. New York, Hafner.

Jackson, Wes. 1991. "Hierarchical Levels: Emergent Qualities, Ecosystems and the Ground for a New Agriculture," pp. 132–53. In W. I. Thompson, *Gaia 2: Emergence, the New Science of Becoming*. Hudson, N.Y., Lindisfarne Press.

Koestler, Arthur. 1978. "The Holarchy," pp. 23–56. In *Janus: A Summing Up*. New York, Random House.

Lubchenko, Jane 1995. "The Relevance of Ecology: The Societal Context and Disciplinary Implications of Linkage across Levels of Ecological Organization," pp. 297–305. In O. G. Jones and J. H. Lawton, eds., *Linking Species and Ecosystems.* New York, Chapman and Hall.

MacMahon, James A., Donald L. Phillips, James V. Robinson, and David Schimpf. 1978. "Levels of Biological Organization: An Organism-Centered Approach." *Bioscience* 28: 700–704.

Novikoff, Alex B. 1945. "The Concept of Integrative Levels and Biology." *Science* 101: 209–15.

Pattee, Howard H. 1973. *Hierarchy Theory: The Challenge of Complex Systems.* New York, Braziller.

Rowen, J. S. 1961. "The Level-of-Integration Concept and Ecology." *Ecology* 42: 420–27.

Thomas, Lewis. 1974. "Systems within Systems." *Harpers* 248: 100–101.

Young, Gerald. 1992. "Between the Atom and the Void: Hierarchy in Human Ecology." *Advances in Human Ecology* 1: 119–47.

Cluster 2

Land and Water Systems

In this cluster we will apply the foundation concepts introduced in the previous cluster to a set of geographically defined ecological systems of land and water. We will construct a hierarchy using the size or scale of system as our ordering principle. Our largest system will be the planet Earth; the smallest will be a pond, a patch of forest, or an agricultural field. It is important to remember that between the large- and small-scale systems many more hierarchical levels can be identified than we will consider here. There is no correct order or correct list of systems that occur in nature.

Once we have a hierarchy of ecological systems, our next goal is to identify processes that link the systems together from top to bottom and across the hierarchy. We will focus on flows of energy, cycles of essential chemicals, nutrients, and water. Energy, water, and nutrients are necessary to maintain life across all the hierarchical levels. The principles found at one level can be applied at other levels, although the details will be different.

Beyond describing the organization and processes of land-water systems, we will also address questions about ecological systems in general. For example, are ecological systems in a state of equilibrium? If so, what could maintain an equilibrium?

4

The Ecosphere, with Comments on the Gaia Hypothesis and the Biosphere

The largest environmental system with which any of us is personally acquainted is the planet Earth. Perhaps the most potent symbol of the Earth system is the photograph of the planet from space (fig. 4.1). The picture of a small round blue, white, and green planet impresses us even years after the picture was made. The picture allows us see Earth as an object in space. No longer do we have to rely on extrapolation from the curve of the horizon or believe in the abstractions of the geographer to tell us the Earth's shape. We can see for ourselves. This one picture may do more to motivate humans to have a global consciousness than all the books and scientific papers written about the global environment.

With this picture in mind, we can diagram the Earth's environmental system in the form introduced in figure 2.2. In figure 4.2, the line around the model represents the boundary of the Earth system, the open space within the box represents everything on or in the earth, the space outside the box represents the solar system, and the arrows leading into and out of the box represent matter and energy flowing into or out of the Earth from or to the solar system. Our first task, which we will carry out in the next chapter, will be to describe the dynamic nature of the couplings of the solar and Earth systems represented by the arrows in the conceptual model. Then we can begin unpacking the Earth system, considering, in turn, various subsystems of environmental interest. In this chapter we will examine various ways in which scientists have expressed their concept of a whole-Earth system.

First, treating the Earth as a discrete physical body implies that it has a boundary. Consideration of the location and character of this boundary raises the problem of definition. The Earth's boundary is a zone extending from the surface of the planet into deep space. In this zone the properties of the atmosphere change continuously with distance. For example, from the

Fig. 4.1   The Earth as seen from space. Photograph courtesy of NASA.

surface to a height of about 100 kilometers the pressure decreases from about 1013 to 0.01 millibar and the air temperature declines from about 15 to −90 degrees centigrade. The part of this boundary zone where life is active is only a few meters high. This is where dust, pollen, flying birds, and bats are encountered. But humans have an impact on the gaseous envelope around the Earth, effecting change in the dust content, the ozone content, carbon dioxide, and so on. Probably, the boundary of the atmosphere should be placed much further from the surface than the point where life is actively present.

Description of the Earth as an ecological system is useful for our purposes because it links the largest system we can know directly with the smallest systems we live among. It is a technical metaphor that enables us to trace flows of energy, cycles of chemical elements, and movements of organisms across many levels of scale. Probably the most useful way to refer

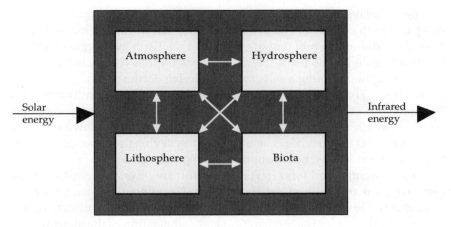

Fig. 4.2 The Earth system described in a systems context.

to the Earth in the system sense is as the ecosphere (Cole 1958). *Ecosphere* emphasizes the spherical nature of the Earth in space and the term is consistent with those relating to environmental systems generally. But it is only one of many metaphors that have been used to describe Earth as a unity. Two of these metaphors originate in science and have had wide usage: the earth as Gaia and as the biosphere. Both metaphors emphasize important components of the ecosphere concept.

The Gaia Hypothesis

A popular concept of the Earth as a whole system is the Gaia Hypothesis of James Lovelock, a chemist, inventor, and independent scientist who has a doctorate in medicine and works out of his home in Cornwall, England. Lovelock invented the electron capture detector, which is in wide use today in gas chromatographs and other scientific instruments. A prominent American supporter of the Gaia Hypothesis is Lynn Margulis, a microbiologist at the University of Massachusetts. The name of the concept refers to the Greek Earth goddess, Gaia.

The Gaia Hypothesis states that "the Earth's surface conditions are regulated by the activities of life." Specifically, the Earth's atmosphere is maintained far from chemical equilibrium with respect to its composition of reactive gases, oxidation-reduction state, alkalinity-acidity, albedo (or reflection from the surface), and temperature by the metabolic activities of the biota. The hypothesis implies that were life to be eliminated, "the surface conditions on Earth would revert to those interpolated for a planet between Mars and Venus" (Margulis and Hinkle 1988).

The Gaia Hypothesis implies that the Earth is a single system, that its unity is due to complex control mechanisms, that the active nexus of this control resides in life, and that life maintains the Earth environment within limits tolerable for life itself. The hypothesis implies that Gaia has a purpose or goal: maintaining a self-regulated, living Earth system.

The Gaia Hypothesis has attracted widespread criticism, especially from physical scientists who claim that physical-chemical processes alone could maintain the Earth environment within reasonably narrow limits. They also point out that life is characterized as having the capacity to adapt to physical conditions, not to control and regulate them.

James Kirchner and John Hart (1988), who are physical scientists, suggest that there are various Gaia Hypotheses: (1) influential Gaia, which asserts that the biota has a substantial influence over certain aspects of the abiotic world, such as the temperature and composition of the atmosphere; (2) coevolutionary Gaia, which asserts that the biota influence their abiotic environment and that the environment, in turn, influences the evolution of the biota by Darwinian processes; (3) homeostatic Gaia, in which the biota influence the abiotic world in a way that enhances its long-term stability; (4) teleological Gaia, which holds that the atmosphere is kept in homeostasis by and for the biota; and (5) optimizing Gaia, which asserts that the biota manipulate their physical environment for the purpose of creating biologically optimum conditions for themselves.

Kirchner and Hart conclude that the first three hypotheses are reasonably well defined, internally consistent, and free of tautology but merely restate older ideas in a Gaian framework, whereas the latter two hypotheses are ill defined and tautological.

Three major points arising from the Gaia Hypothesis are relevant to our interest in ecological systems. First, consistent with our approach, the hypothesis treats the Earth as a single system. Second, the hypothesis emphasizes the role of living organisms in regulating global processes. This point will not be understood until we consider specific processes. But it is obvious that if we aggregate a group of organisms such as bacteria and then consider their collective role in decomposing organic matter globally, converting the carbon in that organic material into carbon dioxide gas that enters the atmosphere, these living organisms have a global role to play. The task is to describe the quantities of carbon being decomposed globally. Third, the Earth is treated in a way indicating that we have a relationship to it as a whole. It is not merely a physical stage on which we act out our life, nor is it merely a source of materials for our use. In one sense it is living. By that I mean that the presence of life makes the Earth different from a lifeless planet lying between Mars and Venus in our solar system. We know of no other planet like Earth in the solar system or in any other galaxy in the universe. Our concern for and care of the planet seems to make common sense.

The Biosphere Concept

The second Earth metaphor comes from the Russian geochemist Vladimir I. Vernadsky (1863–1945), who made important contributions in many areas of science and holds a special place in the history of science in the former Soviet Union. In 1929, while living in Paris, Vernadsky published a book titled *La Biosphère.* The biosphere, in his view, consisted of that part of the Earth where life occurs. It extends many meters into the land, since bacteria have been discovered in deep rock cores and in ocean trenches to a depth of 4.2 kilometers and a temperature of 110 degrees C. (Fyfe 1996), and upward hundred of meters into the atmosphere.

"To solve biological problems by studying only one, in many respects self-contained, organism is impossible," Vernadsky stated. "We know that an organism in a sphere is not a chance agent—it is part of a complex regular order." Vernadsky characterized this order of living beings as having only a minute probability of returning to the same position in space and time. Atoms are in constant motion or migration. They combine to form complex compounds, disintegrate, and recombine in new configurations, moving from the atmosphere to the oceans to the Earth crust in great cycles. "Living matter actively regulates the geochemical migration of atoms. Owing to this role of living matter over the aeons of geological history, the stability of the biosphere is maintained and both living organisms and the entire biosphere evolve." Vernadsky called this peculiar type of constant variation a "dynamic equilibrium" (Baladin 1982, pp. 109–10).

Thus, in Vernadsky's (1945, 1986) conception of the biosphere, stability —in the sense that life and geochemical processes have persisted throughout geological time—is a function of the continual movement of chemical atoms in and through living organisms. Vernadsky emphasized important differences between living organisms and inert matter. "There is a continual stream of atoms passing to and from living organisms into and out of the biosphere. Within the organism a vast and changing number of molecules are produced by processes not otherwise known in the biosphere. Inert bodies change only from outside causes, with the exception of radioactive materials." And "the chemical composition of living bodies is a function of their own properties. The chemical composition of inert bodies is a function of the properties of the medium in which they are formed" (Vernadsky 1986).

The Gaia Hypothesis and the biosphere concept have several points in common. Both emphasize that the Earth is a system characterized by interaction of physical processes and living organisms. The presence of living organisms has transformed the physical processes, causing the Earth to behave in ways different from those expected for a planet lying between Venus and Mars in the solar system. The dynamic interaction of the physical and living parts of the system produces a regulating action, at least at the level of the entire planet, which maintains a form of loose order. Basically, ecological systems are not random configurations of molecules; rather, they follow physical and chemical rules modified by biological activity to produce the complex systems that we observe in nature.

Inevitably, the meaning of Vernadsky's term changed as other authors used it for their own purposes. Now biosphere is widely used to denote the Earth's biota as a whole or, in contrast, the planet Earth as a whole. Nicholas Polunin and Jacques Grinevald (1988) have urged a return to Vernadsky's original usage.

Implications

Unlike most of the ecological systems we will consider in this book, the Earth system is difficult to experience directly. It takes special skill to think abstractly and imagine the Earth as a sphere in space. It is not surprising that past cultures have thought of it as a flat surface, a series of plates, or a hemispherical globe. Visualizing the Earth as a sphere became easier when the first pictures of the planet were received from space. Yet even so, the Earth is too large, too impersonal, and too complex an object for us readily to relate to it emotionally.

Nevertheless, the Earth is our home. Despite many years of searching with radiotelescopes and other astronomical tools, we have found almost no evidence of life elsewhere in the universe. The equivocation in this statement is due to the recent discovery of lifelike material in a meteor originating on Mars. There is ongoing debate about the significance of this observation. But the Earth-system concept teaches that we cannot escape conditions on Earth by traveling to another planet and starting over, as in a science-fiction novel. We are stuck on the Earth, and it is time to think seriously about managing ourselves to avoid further destruction of our home (Golley 1986).

Beginning a study of ecological concepts with the Earth makes sense from a scientific viewpoint, but it also creates a problem for the reader. Individuals often find it difficult to make the leap from their personal concerns to the Earth as a whole. How, you may ask, can I make any difference at all in these large and timeless processes? This is certainly a valid question and we must begin to address it at the outset, when we are thinking about very large systems.

Discussions of environmental responsibility and ethics usually begin with the individual. An environmental ethic depends upon personal experience with the natural world and is an expression of personal responsibility. As individual humans interact with nature, they may grow to understand their connections with and interdependence on nature and incorporate this understanding in their life styles. Arne Naess (1989) points out that we grow from self toward the whole through experience and thoughtful reflection upon that experience. He calls this process self-realization. As I understand Naess, growth starts from a focus on self in self-awareness and progresses to an ever-enlarging appreciation of others, which ultimately may include the whole Earth and the universe.

Naess's approach requires a healthy, objective evaluation of self and a willingness to take charge of our lives. Many exercises and practices can help individuals increase their understanding of and relationship with nature, community, and ecological wholes. These include meditation, prayer, Zen Buddhist exercises, and spiritual practices of Native Americans and others (LaChapelle 1988; Snyder 1990).

But these suggestions are directed toward members of modern cultures who seek to recreate connections and patterns that have been lost. When humans have had to deal with complex systems in the past, we often have relied on interpretation of experience—that is, on wisdom and intuition—to tell us what to do. This is the basis for the conservative approach to living that we find in uneducated farmers in developing countries and in indigenous peoples everywhere. These people have slowly, over generations, developed successful ways to adapt to and manage their environments. They are reluctant to change their practices. They tend to discount the claim of the expert from a so-called advanced country telling them that their future will be better if they employ Western technology and participate in the great systems of world trade. Having little room to make mistakes, they hold the old way to be the safe way.

The conservatism of many indigenous people is understandable. They do not share the modern Western preoccupation with the future and the belief that progress will lead to a better life. They have adapted in such a way that their material and spiritual life and their built environment fit the natural environment. Where they are maladapted, their culture becomes extinct. Their future involves fine-tuning the human environment interactions, not becoming part of massive social and economic change.

Now, one can begin to understand the appeal of technology to Western cultures such as our own. We have gradually lost our capacity to adapt to the natural environment as we have replaced it with a designed and built environment. But this creates a double bind for Western cultures. On one side we adapt to environments built for specific purposes, but on the other side these environments are inadequate to meet all our needs. As the mis-fit between the technology and built environment and the needs of humans becomes greater, we call on more technology to solve the problems created by technology. Always we are assured that the supersolution is almost ready. To obtain it will require only more money and controlling those who would encourage human development in other directions. But our knowl-edge of the world is perennially incomplete and our imagination is inade-quate to create an analog of the natural environment where we are at home.

The built environment was useful when it helped humans adapt to the natural environment. It created an order and stability for humans that was lacking in the natural environment, and it enhanced human well-being. The built environment became a problem when it replaced the natural en-vironment. The solution to the problems created by technology is to un-derstand the environmental limits of human life and to design sustainable environments in which the built and natural environments interact so that both can adapt within their appropriate context. What percentage of the ecosphere must remain outside of human management in order to remain sustainable? The task is to make technology serve adaptation instead of de-stroying it.

Readings

Cole, Lamont. 1958. "The Ecosphere." *Scientific American* 198: 2–7.

Hargrove, Eugene C. 1986. *Beyond Spaceship Earth: Environmental Ethics and the Solar System.* San Francisco, Sierra Club Books.

Hartman, William K. 1984. "Space Exploration and Environmental Issues." *Environ-mental Ethics* 6:227–39.

Henig, Paul M. 1980. "Exobiologists Continue to Search for Life on Other Planets." *BioScience* 30: 9.

Lovelock, James E. 1979. *Gaia: A New Look at Life on Earth.* Oxford, Oxford Uni-versity Press.

Odum, H. T. 1963. "Limits of Remote Ecosystems Containing Man." *American Biol-ogy Teacher* 25, no. 6: 429–43.

Robinson, G. S. 1975. *Living in Outer Space.* Washington, D.C., Public Affairs Press.

Sagan, Carl. 1973. *The Cosmic Connection.* Garden City, N.Y., Anchor Press.

Vernadsky, Vladimir I. 1945. "The Biosphere and the Noosphere." *American Scien-tist* 33, no. 1: 1–12.

Young, Gerald L., and Tom Bartuska. 1974. "Sphere: Term and Concept as an Inte-grative Device Toward Understanding Environmental Unity." *General Systems Yearbook of the Society for General Systems Research* 19: 219–30.

5

Energy Dynamics

In the last chapter the Earth was defined as an ecological system, which means that it is an object with a behavior. Its behavior is defined as the conversion of inputs to the system into outputs into the system's environment. The behavior of the Earth system mainly involves its energy dynamics, or the way it translates energy input into energy output. This is the subject of this chapter.

The Earth ecological system is part of the solar system composed of planets, satellites, comets, and asteroids which move around a star called the sun. The Earth is an open system that receives energy from the sun and radiates infrared energy to the solar system. The solar system is the Earth's environment. Although energy is the primary input to the Earth, material input to the surface, such as meteors, also may occur and were important in the history of the Earth.

Energy Input

The solar energy input to the Earth is about 1.94 calories per square centimeter per minute. A calorie is a measure of energy; one calorie is the amount of heat required, at one atmosphere of pressure, to raise one gram of water one degree centigrade. Although solar energy is the fuel that powers the Earth's surface systems and is the subject of our concern here, other sources of energy on the Earth are derived from radioactive decay and the heat of the core and mantle. The radioactivity was present in the materials that coagulated to form the Earth; the heat in the core and mantle is most likely due to compression.

The Laws of Thermodynamics

The energy dynamics of any system follow the physical laws of thermodynamics. The first law states that energy can be neither created nor destroyed. This means that if we know a system's input or output and the storage of energy inside the system, we can calculate its energy balance. If we know the input, then all of the input energy must exit the system or be stored within it. The first law gives us an energy accounting system.

The second law concerns the energy state of the system. Energy is defined as the capacity to do work. In doing work, energy changes its state or, to put it another way, its capacity to do further work. We can trace the amount of energy flowing through the system, following the first law, but its ability to do work changes even though the amount remains the same. The second law declares that as energy changes state, it becomes less able to do further work. Thus, every step in energy transformation and flow through a system involves the gradual loss of the capacity to do work. This loss is called entropy; as energy is transformed and work is accomplished, entropy increases.

The Character of Solar Radiation

Energy is transmitted to the Earth as packets called quanta. Quanta are often visualized as waves of light. The distance between the peaks of these waves—that is, their wavelengths—varies (fig. 5.1). Humans can see light with wavelengths from 0.000038 to 0.000078 centimeters (or approximately 0.4 to 0.8 micrometers). The amount of energy is related to the frequency (numbers per second) of the waves; the shorter the wavelength, the higher the energy. Short-wave gamma rays and X rays can be very destructive to life and their use is closely regulated. Ultraviolet radiation is the source of sunburn. Long-wave infrared radiation we sense as heat.

The amount of solar energy in the ultraviolet, visible, and infrared wavelengths is shown in figure 5.2. The majority of the energy is in the visible region. The energy tails off on either side of the visible peak. Figure 5.2 also illustrates the effect of clouds, dust, and gases in the atmosphere in intercepting and reflecting incoming solar radiation back into space. Certain wavelengths of energy are completely stopped from reaching the Earth's surface. Figure 5.2 illustrates how changes in the properties of the atmosphere can change the character of the solar radiation received at the surface. This is the physical evidence that underlies the concept of global change in the climate.

Wavelength

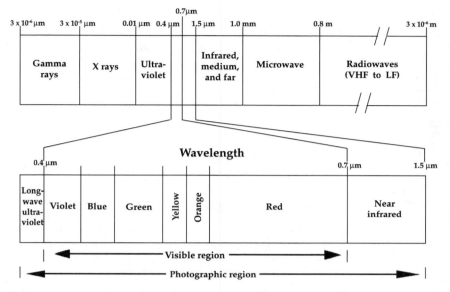

Fig. 5.1   A diagram of the electromagnetic spectrum from the sun received on Earth.

Comparison of the Earth, Venus, and Mars

Comparison of the energy and atmospheric properties of the Earth with those of our neighboring planets, Venus and Mars, will make the unique character of the Earth clearer. Venus is closer to the sun than the Earth—about 108 million kilometers, as compared to 150 million for the Earth (Baugher 1988)—and receives 1.9 times the solar energy of the Earth. Its mass also is slightly less than that of Earth. If the mass of the Earth is 1, then the mass of Venus is 0.815. Its equatorial radius is 6,052 kilometers, compared with the Earth's 6,378. But these are relatively slight differences. Venus is probably closer in size and form to the Earth than any other planet. Mars, on the other hand, is 1 ½ times farther from the sun than the Earth. Although it rotates in the same direction as the Earth, has a similar day-length of about twenty-four hours, and has seasons, it is only about one-tenth as large as the Earth. Mars has no magnetic field.

The atmospheric properties of Venus, however, are totally different from those of the Earth. The Soviet Union's space probes that landed on Venus showed that the surface temperature is about 460 degrees centigrade. This surface temperature neither changes at night nor varies from pole to equator. Venus's atmosphere is 96 percent carbon dioxide and has a layer of

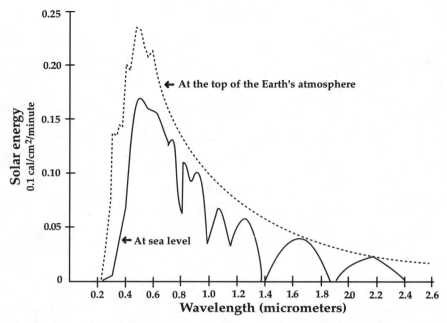

Fig. 5.2   The distribution of solar energy at the outer edge of the earth's atmosphere and on the surface. Energy data in 0.1 cal/cm²/minute.

clouds 15 kilometers thick, starting about 45 kilometers above the surface. These clouds are made of water vapor and sulfuric acid, and they allow only about 1 percent of the solar energy to reach the surface. Some scientists attribute this situation to a runaway greenhouse effect (Degens 1989). According to this hypothesis, higher heat on the Venusian surface baked the carbon from the rocks. The carbon combined with oxygen and entered the atmosphere, and the resulting gas molecules reflected energy emitted by the surface back to the surface. As a consequence, the surface temperature increased, leading to further carbon loss from the surface. The amount of carbon gas on Venus is about equal to the inventory of the Earth's terrestrial carbon, but most carbon on the Earth is stored in sedimentary rock.

Mars has been studied by several American space probes. In television pictures from the 1997 mission, the surface looks like a terrestrial desert, with rock and loose dust and sandlike dunes. Samples of the soil showed that chemically it also was similar to the Earth, with abundant silicon, iron, magnesium, calcium, and other elements. The Martian atmosphere, however, is very different from the Earth's. By volume, it consists of 95.6 percent carbon dioxide and only 0.1 percent oxygen. There is also a large amount of dust in the atmosphere, especially during seasons with dust

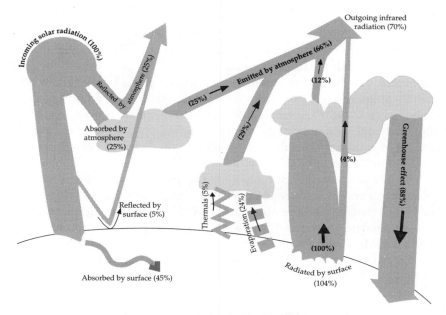

Fig. 5.3 The flux of solar radiation to and from the Earth's surface, as a percentage of the total energy received.

storms. Water vapor is present and water is bound in polar ice caps that expand and contract seasonally. In the northern hemisphere, summer daytime temperatures are −10 degrees centigrade, falling to −85 degrees at night. Winter temperatures may be as low as −125 degrees centigrade.

On Mars the ultraviolet radiation of the sun has slowly destroyed the gases in the atmosphere, so that almost all of the planet's nitrogen, and a large fraction of its oxygen and water vapor, has disappeared. The greenhouse effect of the atmosphere has declined (unlike the situation on Venus), the surface has become colder, and the water supply has frozen.

Obviously, the energy environment of the Earth is very unlike that of Venus and Mars. The Earth's atmosphere is relatively transparent to incoming radiation, with about 25 percent reflected back into space from clouds (fig. 5.3). About 45 percent of the solar radiation is absorbed by the surface, and of this 53 percent is used in the evaporation of water vapor; the remainder is reflected back to space as infrared radiation (Gates 1962). The energy input maintains a physical-chemical environment supportive of life.

The Earth not only transforms energy and does work but it also uses energy to create and maintain higher-order energy structures—living organisms. Solar energy is collected through the process of photosynthesis in green plants and converted into chemical energy stored in organic compounds. According to the second law of thermodynamics, there is a tax of about 50 percent on this conversion process; that is, about half of the energy fixed in photosynthesis is required to operate and maintain the photosynthetic process and is expended as heat energy. As I will discuss later, photosynthesis uses carbon dioxide as a chemical source and produces oxygen as an output. The oxygen-rich Earth atmosphere is partly the result of photosynthesizing organisms. The atmospheric oxygen level of about 20 percent by volume probably was reached in the Paleozoic era, which extended from 250 to 570 million years ago.

The role of organisms is fundamental to understanding the differences between the energy balances of the Earth, Venus, and Mars. Living organisms have transformed the atmosphere of the Earth, the surface, and the rates of deposition and sedimentation, regulating its surface chemistry. Venus and Mars, without life, operate according to physical and chemical principles appropriate to their location in space-time. This activity of organisms is a justification of the Gaia Hypothesis discussed in Chapter 4.

Distribution of Solar Energy

The receipt of solar energy is not equal across the Earth's surface. Because the Earth is a sphere, the amount of solar energy received is largest at the equator and decreases toward the poles (Linton 1965). Further, the oscillation of the sphere causes seasonal variations in energy receipt. In dynamic systems high stocks of resources, such as energy, tend to flow to regions with low stocks unless there are barriers to the flows. In this case, energy input is highest near the equator and lowest at the poles, and therefore energy flows from the equator to the poles. The fact that the Earth is rotating on its axis means that the lines of energy transport are not strictly to the north or south but flow around the Earth through the atmosphere and hydrosphere in the great currents of air or water, creating the climate.

Implications

Our consideration of the energy balance of the Earth has introduced a model that will be used throughout this book. This model depicts a system that is maintained through transformation of inputs into outputs. The particular distribution and abundance of the physical, chemical, and, in the case of the Earth, biological features of the system create the very different conditions we can observe among the planets making up the solar system.

The energy principles introduce another generalization underlying environmental dynamics. I have described the energetics of the Earth as a dualism, energy in and energy out. The equation is balanced according to the first law of thermodynamics, which states that energy in equals energy out. We will observe examples of dynamic balanced relationships in environmental systems at all levels of scale. Principles in physics, chemistry, biology, ecology, economics, and demography usually involve balanced relationships. Deviation from balance attracts our attention immediately and often elicits questions or raises problems that need solutions.

The laws of thermodynamics create a fixed context that is unlike the biological or social context in which we live. The biological and social context is a continual state of change involving adaptation to environmental conditions, including other organisms. Although we seldom visualize biological or social life in thermodynamic terms, it would be helpful to take this view. The laws of thermodynamics set limits on both rates and quantities. They provide an algebra with which we can understand the consequences of our actions.

Consider, for example, the human use of solar energy. Only a small percentage of the energy received by the Earth is captured in photosynthesis and made available to living organisms. Until very recently, humans used a relatively small amount of this transformed energy, even though energy was a limiting factor to human population growth and social development. The discovery of fossil fuels and radioactive energy provided new sources of energy that allowed human populations to expand enormously. These growing populations need organic food and fiber, so some of the available fossil-fuel energy has been used to increase and extend the rate of exploitation of the nonhuman living organisms and ecological communities. P. M. Vitousek and his colleagues (1986) estimate that humans currently direct almost 40 percent of the entire amount of terrestrial fixed carbon to human use or misuse (table 5.1). Athough only 3.5 percent of the terrestrial production is consumed as food or product, 24 percent of the carbon is in communities that are under human management and 12 percent is lost or degraded.

Readings

Gates, D. M. 1985. *Energy and Ecology.* Sunderland, Mass., Sinauer Associates.

Hulbert, M. K. 1971. "The Energy Resources of the Earth." *Scientific American* 224–25, no. 3: 60–70.

Odum, H. T., and E. C. Odum. 1981. *Energy Basis for Man and Nature.* 2d ed. New York, McGraw-Hill.

Table 5.1 Human uses of the products of world photosynthesis. From Vitousek et al. 1986.

	Gt C/yr*	% TOTAL NPP (121 Gt/yr)	% Land NPP** (75 Gt/yr)
Net Plant Production (NPP)			
Present production	112.0		
Lost or degraded by human activity	9.0		
Past production	121.0		
Used Directly			
Food from cultivated land	0.4		
Fodder for domestic animals	1.1		
Wood products	1.1		
Fisheries	1.0		
	3.6	3.0	3.5
Coopted			
(biotic communities altered by human activity)			
Cultivated land	7.1		
Grazing land	4.7		
Forest land	5.7		
Aquatic ecosystems	—		
Occupied by humans	0.2		
	17.7	15.0	24.0
Lost or Degraded			
(production reduced by human activity)			
Forest conversion to agriculture	4.5		
Forest conversion to pasture	0.7		
Desertification	2.3		
Occupied by humans	1.3		
	8.8	7.3	12.0
Grand Total	**30.1**	**25.0**	**39.0**

*Calculated from dry weight, assuming carbon to account for %50. A gt is a gigaton or a billion tons (i.e., 10^9 tons).
**Excluding aquatic production and human use of it.

6

The Composition of the Earth

In the preceding chapter we examined the Earth as a system in which the conversion of energy input into output, following the laws of thermodynamics, was the behavior of interest. Here I will analyze the Earth system by identifying four subsystems: the atmosphere, hydrosphere, lithosphere, and biosphere. We will consider each of these subsystems in turn, focusing on some of the properties that influence their interaction with each other. The unifying theme through all these discussions is the dynamic character of each sphere.

The Atmosphere

In the preceding discussion of the energy dynamics of the planet, I noted that the incoming solar energy was either absorbed by gas molecules in the atmosphere, used in the evaporation of water, or absorbed by the surface and then reradiated as infrared radiation. This means that the immediate surface of the Earth and the atmosphere are the locus of energy receipt and transport. For this reason I will begin my consideration of the subsystems of the Earth system by examining the atmosphere.

The atmosphere is defined as the envelope of gases that extends from the Earth's surface to about one thousand kilometers above the surface. The gases are held to the Earth's surface by gravity. Air pressure and density of gas molecules decrease with altitude. At about seven hundred kilometers above the surface there is almost a perfect vacuum.

Our concern is mainly with the layers of air near the ground, called the troposphere. The troposphere has an altitude of about seven to nine kilometers in the polar regions and about sixteen kilometers in the equatorial regions. In comparison with the lithosphere and hydrosphere, the atmosphere is extremely dynamic. Gases have much less inertia than do water or rock, and they move in short time spans. For this reason mobility is a primary feature of the atmosphere.

Atmospheric air currents are driven by solar energy. The Earth's tilt angle and rate of rotation mean that different parts of the surface receive different amounts of energy. The nature of the surface itself also influences the absorption and reflection of energy. The air in regions of high solar input heats up and the absorbed energy is moved to areas with lower energy input. Differential receipt of solar energy creates the conditions for atmospheric movements. Here again we find that the physical tendency to seek a balance or equality between different energy levels provides a force for dynamic movement.

The general circulation pattern of the atmosphere (fig. 6.1) consists of a series of cells of air movement. At the equator the equatorial convergence zone is characterized by rising air masses that carry a large amount of moisture, producing high rainfall. Circulation of air in the tropical region is from east to west. As the energy is dissipated, the drier air moves in an anticyclonic, easterly direction forming the dry subtropical high-pressure belts (H) that lie north or south of the tropical zone. Further to the north or south in midlatitudes are rain-bearing westerlies (L).

These patterns shift seasonally, annually, and from one year to another. The television weather channel provides a tremendous amount of information on these movements and most of us are familiar with the weather patterns near our home. But local systems are coupled to other air masses, and local weather may be influenced by events thousands of kilometers away.

An example of atmospheric circulation that demonstrates the coupling of air masses and the atmosphere and hydrosphere is the "El Niño" (Christ child) phenomenon. At about Christmastime warm water emerges off the coasts of Chile, Peru, and Ecuador in the stream of the usually cold Humboldt Current, which flows north along the coast of South America. Some years this warming is more intense and it depresses the flow of upwelling cold and nutrient-rich water, decreasing ocean productivity. Fishermen along this coast coined the name El Niño to describe the phenomenon responsible for these annual changes in their harvests.

In the tropical regions there is a convective rising of air from the sea surface warmed by solar insolation. When the sea surface is warm, air may rise ten to twelve kilometers in the troposphere, where it releases its latent heat and then falls. Cool sea-surface temperatures are associated with sinking air masses. In addition to these cells of rising and falling air, there are air movements from east to west that influence water currents in the oceans. For example, easterly trade winds in the Pacific propel ocean currents to the west and create changes in the sea level. Under typical conditions the

North Pole

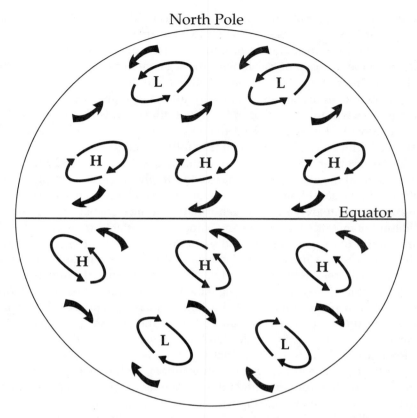

Equator

Fig. 6.1 General circulation of the atmosphere, comparing the westerlies and midlatitude cyclones (*L*) with the subtropical high-pressure belt (*H*).

winds cause a rise of sea level in the west and a decline in the east. If the easterly winds are reduced, sea level in the west falls and a reciprocal change occurs off South America. But this coupled system is not constant; it oscillates with a periodicity of about two to seven years. In an El Niño year atmospheric pressure rises over the Indonesian archipelago and declines along the equator, Pacific trade winds are weaker, sea level falls in the west, and torrential rains may occur in the central and eastern Pacific. The normal patterns of atmospheric movement may even reverse themselves, with catastrophic effect on climate and associated organisms. The 1982–83 El Niño was especially severe, altering weather patterns and causing an estimated ten to twenty billion dollars in damage worldwide.

The chemistry of the atmosphere is dominated by three gases, nitrogen, oxygen, and argon, in a ratio of 78:21:1. Argon is derived by radioactive decay from potassium, which is part of the basic lithosphere (Degens 1989, pp. 196–97). Nitrogen also is thought to be ultimately derived from Earth or lithospheric sources, but at present the atmosphere contains about 90 percent of the nitrogen on earth, with the sea containing 1 percent and the rest in marine sediments. Nitrogen is released to the atmosphere by metamorphism and granitization. Eventually, all the nitrogen will be in the atmosphere except for the nitrogen fixed by living organisms and lightning, which converts dinitrogen (N_2) into fixed nitrogen. The residence time of nitrogen in the atmosphere is about fifty million years.

In contrast to argon and nitrogen, which behave conservatively, the oxygen content of the atmosphere is derived from photosynthesis of plants on land and in the sea. Photosynthesis involves the conversion of carbon dioxide and water into oxygen and organic matter. Some oxygen is consumed in oxidation of metals such as iron and some is sequestered in fossil fuel, but the majority of oxygen is regulated by photosynthesis and organic degradation processes. A higher abundance of oxygen in the atmosphere would trigger forest fires, reducing the plant mass and ultimately lowering the oxygen content. The consequent higher content of carbon dioxide from the destruction of plant biomass would then stimulate plant growth and increase the surface temperature, and increase the amount of oxygen. The oxygen and carbon dioxide cycles tend to be balanced globally.

From an environmental perspective, there is considerable interest in several trace gas constituents of the atmosphere, including carbon dioxide and ozone. Carbon dioxide enters the atmosphere as a by-product of photosynthesis and through decomposition and burning of carbon-rich organic matter. Carbon dioxide levels have increased with industrialization and the use of fossil fuel. Carbon dioxide and other trace gases, including water vapor, lower the rate of thermal infrared radiation from the Earth. That is, these gases allow the passage of sunlight to the Earth's surface but retain some of the reflected light and radiated infrared energy in the lower troposphere, causing a rise in temperature at the surface. Thus, the explanation for global warming is that an increasing level of carbon dioxide in the atmosphere, derived from industry, transportation, and urban sources, acts as a greenhouse gas, trapping infrared radiation in the troposphere and causing its temperature to rise.

Global warming would, in turn, have potential impacts on the production of foods and uncultivated plants and animals, the melt rates of glaciers, and other features of the planet. These changes would generate increased costs and create technical problems. The controversy over the reality of global change has focused on the question of causation. Some have argued that the rising carbon dioxide levels in the atmosphere were an example of natural variation. In 1996, however, several official committees of the United States National Academy of Sciences and the United Nations examined the evidence and concluded that rising CO_2 levels have a human source.

Another trace gas of concern is ozone. Consisting of three oxygen atoms, it is a pollutant in cities, where it may cause health problems. Ordinarily, ozone is found in the stratosphere above the troposphere, where it filters out high-energy ultraviolet radiation that is harmful to life. Ozone also absorbs solar energy; it is one hundred times more efficient than CO_2 in retaining energy. In the stratosphere ozone is generated by dissociation of molecular oxygen by ultraviolet radiation. In the last several decades, scientists have described a reduction in ozone over the South Pole, with the formation of a hole in the ozone layer occurring in the spring. This hole is growing in size annually and the depth of the ozone layer is declining in other regions as well. The consequence of a decline in the ozone layer is increased ultraviolet radiation striking the surface, causing skin cancer in humans and a host of effects on other living organisms.

The depletion of stratospheric ozone is caused in part by chlorine derived from chlorofluorocarbons (CFCs) used for refrigeration and other industrial purposes. Gaseous chlorine reaching the stratosphere reacts with ozone, forming oxygen atoms. The chlorine compounds consume ozone, destroying its protective power. The longevity of CFCs in the troposphere and slow diffusion into the stratosphere mean that it will be many years before ozone destruction would stop, even if CFC production were stopped worldwide.

The Hydrosphere

The hydrosphere consists of the oceans and fresh water. The ocean makes up about 71 percent of the surface area of the Earth. Its mean depth is about thirty-eight hundred meters. Water makes up a larger proportion of the southern hemisphere than of the northern hemisphere.

The physical properties of water are dependent upon its temperature and chemistry. As water temperature decreases, its density increases. Denser water moves to the bottom of a water body under the force of gravity. Water is most dense at 3.9 degrees centigrade. As water temperature declines to 0 degrees, water becomes solid ice. Because ice is less dense (about 8.5 percent lighter) than liquid water, it floats on the surface and the oceans do not freeze from the bottom up. If water continued to increase in density with declining temperature, most of the water on the Earth would be locked up as ice.

The hydrosphere is a dynamic subsystem that is mixed as currents move water from one region to another. For example, the euphotic zone of

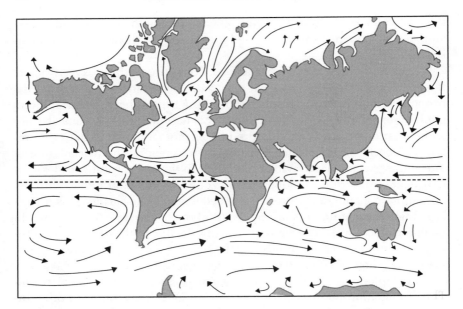

Fig. 6.2   Major ocean surface currents. The arrows indicate direction of flow.

the ocean where living plants occur turns over in about fifty years and the deep-ocean water turns over in about five hundred years (Degens 1989). Two principal processes are involved in these movements. First, winds blow across the water surface and produce surface currents (fig. 6.2). Because of the Earth's rotation and the presence of the continents, these currents do not move in the same direction as those of the atmosphere. Rather, they are deflected to the right in the northern hemisphere and to the left in the southern hemisphere, yielding a clockwise and a counterclockwise movement, respectively. This deflection is called the Coriolis effect, after Gaspard Gustave de Coriolis (1792–1843), who was the first to observe it.

Second, the ocean water is mixed through its currents. For example, in the Atlantic Ocean the water is moved north in the surface current and cools as it reaches the arctic zone. Large volumes of water sink to their appropriate density levels and then return south toward the equator, where they gradually warm and rise. The arctic ocean water carries quantities of oxygen and nutrients, and this oxygenated water ventilates the ocean. These zones of upwelling are present in relatively restricted areas, such as off the coast of Peru. The colder, nutrient-rich deep water wells up and fertilizes the surface so that aquatic plants can grow rapidly, feeding the microscopic consumers and fish. Consequently, these regions contain especially rich fisheries.

Sea water receives chemical inputs from land and rivers, by exchange with the atmosphere, and from deep-ocean vents. Seven major elements occur in sea salt: sodium, magnesium, calcium, potassium, chlorine, sulfur, and carbon. There are thirty-five grams of salt per liter of water. The salinity of the sea is relatively constant over all oceans, supporting the observation that the oceans are mixed. The annual flow of rivers into the oceans represents about forty thousand cubic kilometers of water. Most of the sediment carried by the rivers is deposited near shore in estuaries and coastal plains. Eventually the sand and silt is moved to the ocean by flow of water or wind from land. The clays are attached to living organisms or microscopic flocculent masses and eventually float to the bottom sediment. The accumulation of material from the land may form sedimentary beds, which under pressure and heat form rocks that may be moved to heights through mountain building.

The Lithosphere

The lithosphere consists of the rocks and the inner core of the earth below the point where living organisms are active. The lithosphere is made up of a crust and an upper mantle, which is roughly synonymous with the term *plate.* The plates rest on a partially melted mantle and core, which together are called the asthenosphere. The core is thought to be made up of iron and nickel, with lighter elements such as sulfur and silicon. Apparently the core rotates at a faster rate than the rest of the planet.

The rate of activity of the lithosphere is many orders of magnitude less than those of the hydrosphere and atmosphere. For example, the oldest rocks are about 3.9 billion years old (the Earth itself is about 4.5 billion years old). But under the oceans where the crust is thinnest, the average life span of the rocks is only about 55 million years. If the ocean crust had recycled at that rate for 3.9 billion years, it would have formed and been destroyed only thirty-four times.

If we had been present 300 million years ago, we would have seen a planet with a very different pattern of land and ocean than we see today (fig. 6.3). The terrestrial surface would have formed a single continent, called Pangea. About 200 million years ago this ancient continent began to break up and move in the directions represented by today's configuration of the Earth's surface. Although this movement of the surface took place slowly in human terms, it demonstrates that the Earth is a dynamic, ever-changing system. Not even the solid earth beneath our feet is static and unchanging. The mechanism through which lithospheric change occurs is called plate tectonics.

Geologists have shown that the Earth's surface is made up of plates

Fig. 6.3 A reconstruction of the Earth during the time of the ancient continent Pangaea made by fitting together the major land masses. Pangaea began to break apart about two hundred million years ago.

formed by the upwelling of rock from the deep mantle to the surface (fig. 6.4). The upwelling rock material occurs in midocean ridges, where the expansion of rock causes the plates to move away from the zone of upwelling. The thickness of the plates is determined by temperature. At the 1400 degree centigrade isotherm in the Earth depths, melting transforms the mantle into a plasticlike material. The more rigid crust and upper mantle rides on this lubricated boundary.

Ocean plates have a thickness of about sixty kilometers and continental plates average about one hundred kilometers. The expanding plate moves against the adjacent plate and pushes underneath it into the asthenosphere, where the rock is reincorporated into the magma or slides past it in a transform fault. In some cases reincorporation—technically called subduction— occurs in deep-ocean trenches such as those in the western Pacific. Where subduction occurs, the stress forms volcanoes and uplift of mountains (fig. 6.5). Where plates are moving against one another, there are also earth-

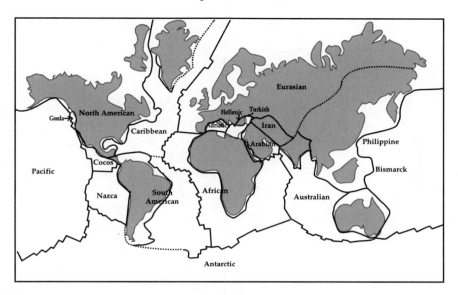

Fig. 6.4   The locations of the major tectonic plates.

quakes. Thus, topographic relief, which is such an important feature of the Earth, is created and maintained through plate tectonics.

The Biosphere

The biosphere consists of all the living organisms that occur on Earth. About 1.7 million species of plants, animals, and microorganisms have been described by scientists and placed in taxonomic lists. The actual total, however, is probably much larger. Estimates range from ten to forty million species.

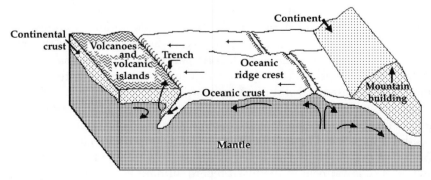

Fig. 6.5   A schematic drawing of the development of a tectonic plate midocean ridge, with movement toward deep-ocean trenches or the continents.

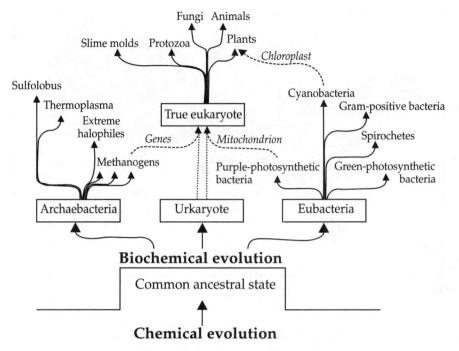

Fig. 6.6 Biochemical evolution from a single ancestral state.

The age of the Earth has been calculated to be about 4.5 billion years. Egon Degens (1989) states that by 4 billion years the planet could have contained environments that today shelter aerobic and anaerobic forms of life. These include hydrothermic hot springs, rivers, lakes, and deep-ocean environments. A period of chemical evolution preceded biochemical evolution leading to living cells (fig. 6.6). Simple organisms are thought to have evolved in three lines: the archaebacteria, which live in extreme environments; the eubacteri, which gave rise to spirochetes and green-photosynthetic bacteria; and the eukaryotes, which led to plants and animals. All share the same genetic code and have similar metabolic pathways. The process of development of new species can be described using a model of a tree, where branching represents evolution of new forms.

After life appeared 3.45 billion years ago, it remained in the relatively simple form described above until about 800 million years ago, when multicellular organisms appeared. Then, about 550 million years ago, in the Cambrian period, there was an explosion of new forms of life. The phyla that exist today appear in the fossil record without evidence of slow, long periods of development leading to them. It is as if the tree of life developed a proliferation of branches, which thereafter branched into twigs. After this dramatic change, diversity slowly increased over the next 100 million years. Then a steady state was reached, followed by a series of extinction events and recovery. There have been at least six of these events in the Earth's history when the number of kinds of organisms was drastically reduced.

If we view these shifts in the biosphere in the context of changes in the other subsystems, they are easier to understand. The biosphere is fundamentally different from the nonliving world because life continuously creates new structures and processes. This creativity is the result of the genetic mechanisms characteristic of all living organisms. Variety comes from such genetic processes as recombination, chromosomal rearrangements, and mutation. These genetic properties give cells information on how to function and direct how the body is built.

The result of genetic variability and sexual reproduction is that within a population each individual has slightly different capacities to respond to the environment. As individuals interact with each other and the environment, some are better able to survive and reproduce than others. The next generation comes from the survivors, and their genetic information is available to be recombined and used to instruct yet another population of individuals.

This creative force has produced the biotic variety that we see on the Earth. I think it is the fundamental reason why life continues to exist on the planet. Even under the stress of the catastrophe of human-caused extinction, life continues to respond by evolution of appropriate forms. If the larger organisms are driven to extinction, then the smaller viruses, bacteria, and protozoa, which humans may classify as disease-promoting organisms, evolve and flourish.

Since the biota must respond to an environment that is dynamic and ever-changing, it requires a mechanism to assure survival under any environmental conditions. It would be impossible to have life somehow anticipate all environmental change and produce in advance the types of organisms that could survive the changing conditions. This is the argument for a designed world, and it no longer fits as an explanation of the complex world we know through biology. In contrast, by constantly producing more variation than can survive, life is so diverse that no environmental change has been sufficiently great to cause total extinction. In a system-science sense, this is an example of Ashby's Law of Requisite Variety, which states that survival of a system requires that it be designed to have a greater capacity for change than the processes of the environment that affect it.

A second important feature of the biosphere is its capacity to respond to and alter the environment. I have observed that the environmental conditions created by the lithosphere, hydrosphere, and atmosphere are distributed in complex patterns across the surface of the Earth. For example, consider the distribution of water across a landscape, with a wet area at a topographic low point and a dry area on the ridge top. Species are distributed across this gradient of moisture, depending upon their genetically defined capacity to live in wet or dry environments. In this case, water or moisture is a limiting factor to the biota. Resource gradients might be divided by many species. Each species responds to limiting factors in its own unique way.

Beside the impact of the environment on the organisms, the reciprocal relationship is also important. Organisms affect the atmosphere, lithosphere, and hydrosphere physically and chemically. For example, the primitive atmosphere of the planet was quite different than it is today. As discussed above, it may have been anoxic and highly reducing chemically. The rise in free oxygen in the air and in the sea, which makes modern life possible, derives from plant photosynthesis. In the photosynthetic process, carbon dioxide and water are converted into simple sugars and oxygen. Degens (1989) estimates that it would take green plants about two thousand years to generate one volume of atmospheric oxygen. It is thought that the present level of oxygen occurred during the Paleozoic era and has stayed around 20 volume percent ever since.

A lifeless planet is subject to physical and chemical forces of dissolution of the surface. Pictures of the surface of the moon or Mars illustrate these conditions. On Earth, vegetation covers the surface, except in the most extreme environments, and modifies the geomorphic forces. Where life exists soil is formed, and with the presence of soil we observe a new ecological reality that combines the physical and chemical elements of all the spheres with life. Soil does not exist unless all of these parts are present.

The consequence of these reciprocal interactions is an Earth system of enormous diversity. From every perspective the regions of the planet differ in important ways. Further, all of the subsystems are dynamic and responsive. They are continually changing as they interact with each other. Although from the perspective of the human organism, which lives about seventy to eighty years and is of middle size, the planet is stable and constant, this impression could not be more fallacious. The process of balancing never ends, but a point of balance is also never achieved. The cycles of change go on endlessly.

Implications: Interaction Between the Spheres

The four divisions of the Earth are useful for analytical purposes, but their environmental interest lies in their interactions and the way they modify the characteristics of each other. Thus, environmental insight derives from the interaction of atmosphere and lithosphere, not from deeper analysis of the atmosphere separate from the lithosphere. Such analysis is the province of the atmospheric physicist and the geologist. The environmental focus brings us back to the concept of the ecosphere as a whole.

An example of the interaction of the subsystems is the explosion of a volcano, such as Pinatubo in the Philippines, which is a consequence of plate tectonics. A volcanic eruption may throw dust into the atmosphere in sufficient quantity that, when it is distributed globally by the process of atmospheric circulation, it reflects incoming solar energy away from the Earth and reduces plant photosynthesis. Some paleontologists think that such an event or the dust caused by a meteor striking the Earth could have caused one of the known extinction events of the biota, such as the destruction of the dinosaurs.

Discussion about the human impact on the Earth and its effect on the global climate is intense and stimulates violent political disagreement over public policy. This concern needs to be set into context. Of course humans have an impact on the four spheres. Human activity is sufficiently intense that the patterns have been shifted in many direct and indirect ways. For example, removal of forest has changed the reflectance of the surface, called the albedo, and changed the absorbance ratio of solar energy, changing climates and influencing the drought cycle. The question is not whether humans have had an impact on the Earth system, but rather whether their impact has increased chaotic behavior, making it more difficult to predict the future and increasing the likelihood of species extinction.

To answer these questions requires a distance in space-time that I mentioned above. But it may be impossible to obtain sufficient distance from Earth to observe ourselves objectively. It is more likely that humans will take various sides on these questions and then argue about other issues that concern them. Today we observe politicians using these issues to advance their own ideologies. I suspect that it will be no different in the future. Global change will continue and we will have increasingly accurate ways to describe what has happened. Presumably, these models will be helpful in predicting patterns while remaining sufficiently unreliable that people who will be affected by changing policies will continue to demand that nothing be done until better models or more information are available.

Readings

Bryson, Reid A. 1974. "A Perspective on Climatic Change." *Science* 184: 753–60.
Changnon, Stanley A., Jr. 1968. "The La Porte Weather Anomaly: Fact or Fiction?" *Bulletin of the American Meteorological Society* 49, no. 1: 4–11.

Howell, David G. 1995. *Principles of Terrane Analysis: New Applications for Global Tectonics.* 2d ed. London, Chapman and Hall.

Langenbein, W. B., and S. A. Schumm. 1958. "Yield of Sediment in Relation to Mean Annual Precipitation." *Transactions of the American Geophysical Union* 39, no. 6: 1076–84.

Man and the Ecosphere. 1971. Readings from *Scientific American.* San Francisco, W. H. Freeman.

Plass, Gilbert N. 1959. "Carbon Dioxide and Climate." *Scientific American* 199: 1–9.

Rosswall, T., R. G. Woodmansee, and P. G. Risser. 1988. *Scales and Global Change: Spatial and Temporal Variability in Biospheric and Geospheric Processes.* Chichester, England, John Wiley.

7

The Biome

I began my analysis of the ecosphere by treating it as a system, following the definition in Chapter 2. The behavior of the ecosphere was expressed in its energy dynamics and my analysis focused on how the subsystems of the lithosphere, hydrosphere, atmosphere, and biosphere interacted to influence the energy dynamics of the system. In this way we created a view of the Earth as an ecological whole or an ecosystem. The Earth itself is a subsystem of the solar system.

This viewpoint is satisfactory as far as it goes. It provides us with an understanding of global processes that have wide impact across many scales. Plate tectonics, climate and climate change, ocean currents, El Niño events, the evolutionary response of organisms are all operative in environmental systems at the global level. But the four spheres or subsystems are not nested systems within the ecosphere, because they do not conform to the rule of self-similarity explained in Chapter 3. For this reason, we will not consider the lithosphere at smaller and smaller scales; that is the subject of geology. Nor we will subdivide the hydrosphere into smaller and smaller scales; that is the task of the hydrologist.

To create a hierarchy of nested systems, each level must be defined by the same features used in defining the highest-order system. The nested set that we need for understanding the environmental patterns of the Earth and for environmental management begins with the ecosphere and ends with the small systems in which we live. If we think of this smallest system as a prism (fig. 7.1), it will have a surface where we live, air above the surface, and soil and rock below the surface. We share the prism with numerous organisms. If the system is aquatic, then water is interposed between the soil and air.

This prism contains parts that are analogous to the lithosphere (the soil

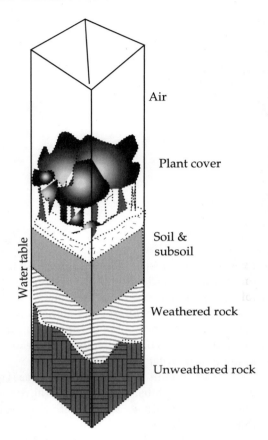

Air

Plant cover

Soil & subsoil

Weathered rock

Unweathered rock

Water table

Fig. 7.1 A landscape prism showing the relationships between physical and biotic components. Based on Fortesque 1980.

and rocks), the atmosphere (the air), the hydrosphere (water), and the biosphere (the resident organisms) which interact to form the system. The analogy means that the prism and the ecosphere are self-similar. They are defined using the same subsystems and system behaviors. Between the prism representing the smallest environment and the Earth representing the largest, there is a hierarchy of other systems. The number of members of the nested set is not fixed. Rather, it depends upon the purpose of the analysis. In this book we will recognize four levels beneath the ecosphere: the biome, the landscape, the watershed, and the ecotope. These terms will be defined in this and the following chapters.

To create our first level of subsystems in the nested set of Earth systems, we have to resort once again to our imaginary spaceship. If we examined Earth from a distance, we would see that the terrestrial surface was occupied by vegetation of different types and statures and that the hydrosphere was covered with water of different colors. These differences in appearance indicate differences in the structure and function of the subsystems. For example, beginning at the North Pole, we observe mainly rock, gravel, ice, and snow, with living organisms reduced to a few mosses, algae, and grasses living in sheltered places among stones. Further south, the surface changes. It is covered by short grass, with mosses and herbs and an occasional willow shrub. This region is called tundra. As we move toward the equator, individual stunted spruce and fir trees appear in the tundra, and eventually we observe full forests dominated by these two genera. Spruce and fir forests grow so tightly that there is deep shade beneath them and the tundra vegetation cannot persist. Yet further to the south, deciduous trees replace the conifers, except in harsh environments. In the center of the continents, however, the aridity is such that trees cannot grow except along rivers. In this region we observe the steppe, or grasslands. Further toward the equator we reach the point where the tropical atmospheric cells have lost their moisture and are sinking to the surface. Their dry condition produces a belt of deserts on the land. Finally, as we reach the equator we observe the tropical rain forest, which consists of as many as three layers of tree canopies, one above the other.

In this imaginary trip we recognized different kinds of plants, which were characterized mainly as having different stature. The technical term for this characteristic is *life form*. The life form of the vegetation changed from the pole to the equator. If we had identified the plants making up the vegetation, we would have noted that the variety and abundance of plants also increased toward the equator. These changes in life form and biodiversity are directly related to the presence of moisture and the temperature. Plants living in the arctic environment may have only six weeks to accomplish their annual growth. Their sensitive buds or growing tips must be protected from the intense cold of winter. In the harsh climate they are incapable of producing sufficient biomass to reach a tall stature. Lack of water is another important limiting factor to plant development and growth. In the most intense desert regions, living organisms are confined to cracks in rocks and other rare habitats where they can persist until a rain falls. At this moment the desert becomes alive, plants and animals go through their life cycle, reproduce and die or return to a hibernating state until the next rain event.

At the coarsest level of analysis there is little disagreement over the identification of the biomes. However, innumerable criteria could be used to make a finer analysis, and each geographer has a slightly different approach. Some display on a computer the climate patterns that control the presence or absence of vegetation and predict what biomes will occur. Oth-

ers use satellite photos of the Earth's surface to produce maps of the biomes. Recalling that any interpretation imposes our perception on a continuum of changing environment, we can choose the approach that meets our
particular purpose. I will use the recent analysis of Robert Bailey (1995), a
geographer who is working for the United States Forest Service, but there
are many other options (Eyre 1971, Walter 1973, Udvardy 1975).

The names we use to describe these systems also vary among geogra-
phers. The subsystems have been called ecoregions, provinces, biomes, and
so on. The term *biome* was coined by the American ecologist Frederic
Clements (Clements 1916) and was meant to include plants, animals, and
environment. It also was used in the International Biological Program
(1965–1972) as a context in which studies of specific ecosystems fit.

Biome Characteristics

I have characterized the biomes by life form of the dominant vegetation.
Since the vegetation has gone through natural selection for the particular
climatic and geologic conditions of the biome, the vegetation is a suitable
indicator of the biome. In that sense, the vegetation integrates the various
characteristics used to identify a biome.

Table 7.1 brings together data on the area of the Earth covered by each
biome, the standing stock of organic matter, and its productivity. We will
examine production in a later chapter, but here it is a useful measure of the
activity of each system. Productivity is a surrogate for energy flow because
it is a measure of the conversion of solar energy by photosynthesis into organic matter. The energy value of plant tissue is about 4.5 kilogram calories
per gram of dry tissue, and one can use this factor to convert these productivity figures into energy equivalents. The area of the biomes is roughly
similar except for the tundra, which covers the least area, and the coniferous forest. The productivity rates can be contrasted to the standing stock
of organic matter stored in the biome. The largest quantities are in the rain
forest, correlated with the high productivity. But the second-highest storage
is in the coniferous forest, which has a substantially lower production rate.
In this case, the low temperatures, high moisture levels, and the acidic nature of the litter of the conifer trees combine to make decomposition of organic matter occur at slow rates, leading to a build-up of humus. The standing stocks in the oceans are very small in comparison with land vegetation
and are not shown on the table. Of course, these figures are for the potential
vegetation of the Earth. Humans have cut down many of the forests and the
amount remaining is less than is shown in the table.

The productivity rates of the biome are controlled by the length of the
growing season and the available moisture and nutrients. The highest an-

Table 7.1 Comparison of area, rate of production, and biomass of the biomes. Data from Emanuel et al. 1985 and Lieth 1975.

Biome	Area 10⁶ Km²	Annual Rate of Production T/Ha	Stock of Organic Matter
Tundra	4.2	0.14	0.2
Coniferous forest	17.4	0.50	3.5
Temperate forest	27.1	1.00	2.4
Grassland	22.9	0.7	0.5
Desert	26.6	0.07	0.1
Tropical forest	31.0	2.0	4.5
Open ocean	326.0	0.1	——*
Coastal zone	36.0	0.2	——
Upwelling areas	3.6	0.6	——

*The stock of biomass is negligible compared to land-based systems.

nual rates are observed in tropical forests. The oceans have much lower rates than terrestrial biomes but occupy a larger area, so that their overall contribution is significant.

Heinrich Walter (1973) presents a gradient of climate, vegetation, and soil across the eastern European lowlands which is illustrative of the biome patterns we are exploring. This gradient (fig. 7.2) goes from tundra to desert. It is only missing the tropical forest to replicate the listing in table 7.1. Rainfall is maximum in the center of the transect, reaching 450 millimeters. At this point the potential evaporation is less than the rainfall, but evaporation increases with temperature until it is almost ten times greater in the desert. The growing season increases from 90 to 230 days and irradiation increases from 30 to 90 kilogram calories per centimeter squared per year. Mean annual temperature ranges from −9 degrees centigrade to +11. Humus layers are deeper under the forest steppe and steppe biomes. Groundwater depth also increases from tundra to desert. The patterns documented in this figure explain the productivity patterns in table 7.1. Tundra organisms must adapt to low temperatures, short growing seasons, and relatively low moisture in comparison with the conditions in forest and grassland biomes.

These relationships have been shown in a different way on the North American continent by Ronald Neilson, an ecologist with the Environmental Protection Agency in Corvallis, Oregon. Neilson (1987) has examined the relationships between the boundaries of these biomes and climate by modeling precipitation and stream runoff on two transects across the United States (fig. 7.3). Figure 7.4 describes the close relationship between vegetation pattern and climate. For example, the diagram showing precipitation each month on a transect from South Carolina to California demon-

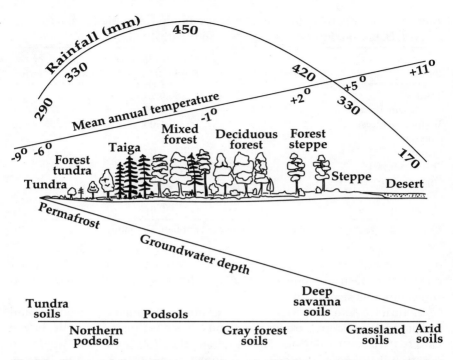

Fig. 7.2 Diagram of climatic factors, vegetation, and soil in the eastern European lowlands. From Walter and Box 1976.

strates a shift from winter precipitation in the east, to summer rainfall in the grassland, to autumn rainfall in the desert. High rainfall in the eastern forest distinguishes it sharply from the grassland and the desert. These diagrams are further evidence that climatic patterns across the continent can be defined spatially and are correlated with vegetation life form and then can be used in mathematical models to predict a variety of environmental phenomena, including productivity.

Interbiome Connections

The biomes are connected by flows of materials and organisms to form the ecosphere. Flows require a mobile medium, such as air or water, which can move in response to energy gradients and under the influence of gravity. The rivers of the world collect the products of continental downwasting and transport them to the oceans, forming marine sediments. The winds transport gaseous products of photosynthesis and decomposition and solid

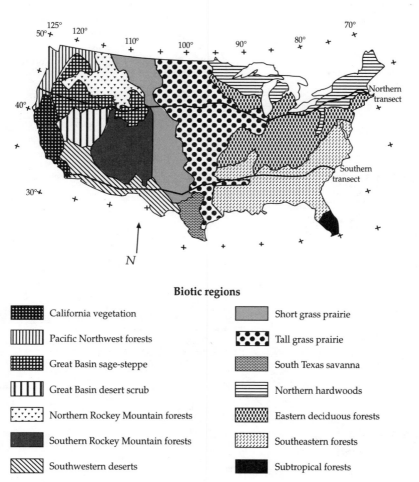

Biotic regions

California vegetation		Short grass prairie	
Pacific Northwest forests		Tall grass prairie	
Great Basin sage-steppe		South Texas savanna	
Great Basin desert scrub		Northern hardwoods	
Northern Rockey Mountain forests		Eastern deciduous forests	
Southern Rockey Mountain forests		Southeastern forests	
Southwestern deserts		Subtropical forests	

Fig. 7.3 A map of the United States showing the transects used by Neilson (1987) to correlate climate patterns with biotic regions.

particles of dust and soil that can form deep layers of soil, called loess. Loess soils are some of the most fertile for agriculture on the Earth.

Further, the biomes are connected through the movements of organisms. Microorganisms and plant seeds and fruits usually are distributed passively by air or water. Many animals move by their own volition and some make regular trips from biome to biome in their migratory movements. These long-distance migrants carry the chemical elements (including pollutants and radioactive materials) obtained in one location and their parasites and disease organisms to another part of the world. The movements of modern

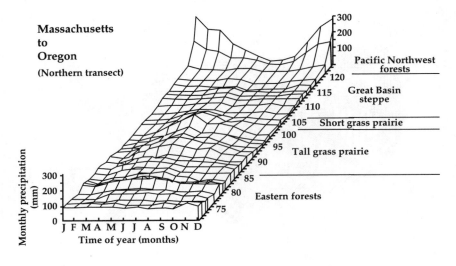

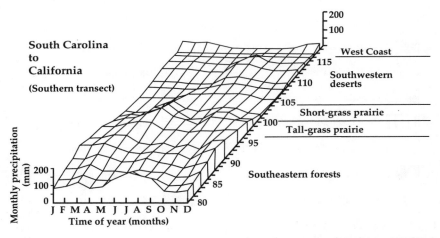

Fig. 7.4 The pattern of monthly precipitation on the southern and northern transects of Neilson 1987.

humans increase the spread of these organisms worldwide—witness the expansion of the AIDS virus, malaria parasite, and cholera. Alfred Crosby (1986) points out how expansion of Europeans, beginning in about A.D. 900, with their fellow travelers, the rat, brome grass, and influenza, changed the biology of entire regions. There are few barriers separating the biomes; the Earth is an integrated ecological system.

Implications

In considering the implications of the biome concept in understanding the environment, it is useful to return to our discussion of the recognition of objects in nature. The reduction of the diversity of nature into patterns that can be understood by the human mind introduces a fundamental problem that will be of concern throughout this book. For example, biomes can be located on another mental construct, a world map. The area occupied by each biome can be calculated (see table 7.1) and then the area of a biome in square meters can be multiplied by the carbon dioxide evolution per square meter. In this way we can calculate the carbon exchange between the atmosphere and the terrestrial and oceanic surface, and then make judgments about the importance of human-caused carbon exchange. Because biomes differ in their dynamic properties, it is important that we have consistent ways to identify them. If we cannot determine the boundaries of an object unambiguously, does the object exist in nature? Or is nature made up of gradients, not objects?

This problem is commonplace in environmental studies. Take the case of an individual organism, which for most of us is unambiguously an object. The organism is exchanging energy, matter, and information with its environment continuously, as long as it is alive. In this sense it is a system of exchange, which involves both the organism and its immediate environment. From the environmental perspective, it is inconceivable that an organism can be considered without its environment. It is entirely dependent upon its environment, and if it were separated in some unimaginable way, it would immediately cease to exist. The two parts create a single unity, an ecosystem. Is an organism an object or not? Or every time we interact with an organism, must we include its environment in our sphere of concern?

The way we view the natural world is culturally determined. Euro-American culture tends to see nature as made up of objects that can be studied, manipulated, and owned. Natural objects are analogous to parts of machines. Machine parts are interchangeable among machines. A machine consists of its parts. This culturally defined perspective leads us to see forests as made up of trees that have monetary value, and to discount all the other properties of forests. It permits us to express highly valued cultural attributes, including accumulation of money and expression of power over nature and other humans, without recognizing the destruction of other properties and values of systems.

The slightest shift in purpose or goal, however, causes us to see differently. In an ecological perspective, natural organisms are unique products of natural selection, evolution, and ecological interaction. The individual organism is an object of scientific and social interest. Yet individual humans viewed as individuals from this perspective are also members of a family, another object viewed from another perspective. What is real and consistent in this analysis is the linkage of apparent objects into complex systems of interactions. Even in the most extreme forms of isolation of objects, our interpretation is almost always confounded by unexpected connections and feedbacks.

From a utilitarian point of view we may emphasize either the objectlike features of nature or the connectedness of nature, depending on our purpose and need. The approach developed in the Foundation Cluster of chapters is designed to help us visualize these patterns without falling into confusion or rejecting the environmental viewpoint entirely. By moving back and forth across these levels of explanation, the environmentalist may focus on individual objects, wholes, and linkages. A complete explanation involves all of these perspectives.

Readings

Bailey, Robert G. 1995. *Ecosystem Geography*. New York, Springer.

Lieth, Helmut, and Robert H. Whittaker. 1975. *Primary Productivity of the Biosphere*. New York, Springer.

Walter, Heinrich. 1973. *Vegetation of the Earth: In Relation to Climate and the Ecophysiological Conditions*. New York, Springer.

Whittaker, R. H. 1970. *Communities and Ecosystems*. New York, Macmillan.

8

The Landscape

The biome is a large environmental system that may be divided into portions occupying different continents. For example, the tundra biome occurs in North America, Eurasia, and Antarctica. Because of size and location, it is likely that biomes are heterogeneous, consisting of many subdivisions. I will refer to these subsystems of the biome as landscapes. In the emerging field of landscape ecology, the term *landscape* is used in many ways. For our purposes, however, landscape systems occupy tens, hundreds, or thousands of square kilometers, as compared to biomes, which occupy millions of square kilometers.

An example will help clarify the concept. In the southern end of the temperate forest biome in North America there are several landscapes which are distinctive and of great practical importance to human commerce. Near the ocean is a relatively flat region that was once covered by water. Its soils are sandy and the vegetation is generally pine on the uplands and swamp forest along rivers and in swamps. This landscape, extending from Maryland to Florida, is called the Atlantic Coastal Plain. It is bordered on the northwest by a different landscape, the Southern Piedmont. This region is geologically old. The surface soils have been destroyed through agricultural exploitation and the red soil that characterizes the region is a weathered metamorphic rock, saprolite, that overlies the parent material. The Piedmont is further characterized by a vegetation with many species, especially oaks, hickories, and upland pine. It is closely related to the Appalachian region, which borders it on the northwest. Each of these divisions has a distinct history and provides different resources, opportunities, and problems. Each is also heterogeneous, with many subdivisions. But the overlying character of each is sufficiently different that any observer can recognize the landscape division.

Analysis of the Landscape

We can begin our analysis of landscapes by considering their structure and function and how these attributes change under environmental influence. This represents the conventional systems approach. Alternatively, we can examine the components of the lithosphere, hydrosphere, atmosphere, and biosphere that are appropriate to a site and consider how these components interact to produce the observed structural and functional attributes of the landscape. I will use both of these approaches to elucidate the landscape concept.

If we view a landscape from an airplane or from aerial photographs, we see a pattern made up of repeating elements. The pattern characterizes each landscape because it represents the interaction of soil, water, air, organisms, and human influences. If the human presence is not dominant—as in parts of the Amazon forest or the Sahara Desert—the interactions of the soil, water, and organisms are naturally expressed. The vegetation scientist calls this condition the potential vegetation. Usually it is difficult to reconstruct the potential vegetation because it has been so changed by humans. As mentioned in the Introduction, Akira Miyawaki has been able to recreate the potential vegetation of Japan (Miyawaki 1980–1989) because the natural forest has been preserved around the shrines and temples. Elsewhere, the Japanese landscape has been highly modified by humans. Miyawaki used these remnants to construct a simulation of the original forest cover and then compared the potential to the actual vegetation patterns to show the extent of this change.

The mosaic of landscape elements can be described as patches on a background matrix of the dominant or the historical vegetation, connected by a complex network of corridors (fig. 8.1). The landscape mosaic (Forman 1995) then forms the structural representation of the landscape system.

Landscape Change

Landscape patterns are not static; they change as the biota responds to changing environmental conditions. Because of human influence, landscape patterns may change relatively rapidly. In the state of Georgia, Monica G. Turner and C. Lynn Ruscher (1988) compared the composition of landscapes in 1930 and in 1980 (table 8.1). They found that the proportion of landscape elements in the mosaic had changed. These changes were caused by abandonment of agricultural land and its natural revegetation by forest, especially in the Piedmont. Where agriculture continued, small crop fields were consolidated into larger fields. Land abandonment was partly due to the deadly effect of the boll weevil on cotton and to the Great De-

Fig. 8.1 Patches and corridors in an agricultural landscape matrix. Photograph by Richard Westmacott.

pression. Land that was changing from farm field to forest was labeled "transitional" in their analysis. Transitional land made up the largest proportion of the landscape in 1930 but had declined significantly by 1980.

The patterns of change differed among the landscapes of Georgia. For example, on the Coastal Plain the average size of agricultural patches increased from 1930 to 1980, while they decreased on the Piedmont. These shifts are explained by the fact that the concentration of agricultural activity shifted from the Piedmont to the Coastal Plain in this fifty-year period, and the use of tractors and machinery required larger fields. Turner and Ruscher concluded that the Georgian landscape had become less fragmented and more connected—a positive change from the perspective of wildlife and conservation.

Similar processes of landscape change are observed everywhere. In the former Czechoslovakia I observed finely divided fields that had been consolidated into very large fields on state farms so that modern equipment

Table 8.1   Proportion (%) of the piedmont and coastal plain landscapes of Georgia occupied by different land uses in 1930 and 1980. Based on selected data from Turner and Rausch 1988.

Land Use	Piedmont		Coastal Plain	
	1930	*1980*	*1930*	*1980*
Urban	1	3	1	1
Agricultural	30	12	25	36
Transitional	51	24	51	12
Coniferous forest	5	41	8	37
Deciduous forest	11	17	12	10

could be used. These changes were made without considering the detail of the landscape mosaic created by centuries of human adaptation. The consequences were serious problems of soil erosion and declines in yield. On the other hand, land abandonment to agriculture is occurring in many parts of Europe under the shift in policy of the European Community and could have positive effects on wildlife and plant conservation (Baudry 1991).

Landscape Processes

The field of landscape ecology has received attention from ecologists partly because it provides a context for the study of large-scale processes. How, these ecologists ask, does landscape pattern control process? Landscapes at the scale of the river basin control flows of water, while landscapes within airsheds control or at least influence the flow of air and the local climate at the ground surface.

An example of how pattern influences process has been described by Lech Ryszkowski and André Kedziora with reference to the landscape of western Poland. This region was covered by glaciation and contains many small lakes and ponds on a flat surface with low hills (called drumlins and eskers). The region has been managed for agriculture for hundreds, if not thousands, of years. At the time of the study the landscape was covered by a variety of crops in large fields of collective farms. The crops included beets, wheat, rape seed, and meadows. Forests, the original cover on the landscape, occurred only in small patches and in shelter belts on the edge of fields and roads. Ryszkowski and Kedziora (1987) measured the flow of energy among these landscape units. Employing the same concepts that I used in describing the heat balance of the Earth (Chapter 5), these researchers determined the amount of energy received and used in each landscape unit independently and as a whole.

Table 8.2 Energy flow in agricultural ecosystems in a Polish landscape. The data are in millijoules per square meter and represent growing season averages. From Ryszkowski and Kedziora 1987.

Energy Input and Use	Ecosystems					
	Shelterbelt	Meadow	Rape Seed	Beets	Wheat	Bare Soil
Solar Input	1730	1494	1551	1536	1536	1575
Evaporation	1522	1250	1163	1136	1090	866
Heating Air	122	215	327	339	385	651
Heating Soil	87	29	61	61	61	47

*Data expressed in mJ per square meter represent means for the growing season.

They found that the net solar radiation input was similar across all landscape units, but that there were wide differences in the way this energy was used. Shelterbelts used more energy for evaporation of water and nearly 5.5 times less energy in heating the air than did bare soil (table 8.2). The reason for these differences is that the trees in the shelterbelt have longer roots than crop plants and can absorb water from deeper soil layers. The meadow ecotope also used more energy in evaporation and less in heating air than did the crops.

These data on energy balance of separate landscape elements cannot be added together to represent the heat balance of the western Poland landscape, because there are interactions between the elements. For example, energy flows from the meadow to the crop field through convective air movement. The energy in the heated air is used by the crop for evaporation of water, which increases the flow of nutrients from soil to plant and enhances plant production. The consequence of these flows between landscape elements is that the energy environment of the landscape is less extreme overall. Ryszkowski and Kedziora point out that optimum production requires an optimum landscape pattern, as well as suitable local environmental conditions of water, nutrients, and solar radiation.

Human Management of Landscape Pattern

Human activity changes landscape patterns and creates a human-dominated landscape. The challenge of land management is to avoid creating destructive landscape patterns accidentally through ignorance of the significance of the impact of landscape pattern on process. An example of this problem can be observed in the western United States, where relatively large areas of mountain forest were uncut until the last few decades. In the Northwest a conifer, the Douglas fir (*Pseudotsuga menziesii*), forms stands of trees that are unusually tall, with a large biomass. Douglas fir is a valuable timber tree and the western Douglas fir forest landscapes have been clear-cut almost throughout its range. Foresters think that Douglas fir reseeds itself best in open, sunlit environments and that clear-cutting the forest is the most efficient method of regenerating the species. But the practice of clear-cutting is directed solely at the Douglas fir. Although the foresters may well be correct, the other species in an old-growth forest will not fare as well as the dominant tree species under this management.

This topic has been highly controversial in the northwest region of North America. But the argument has tended to be restricted to the biology of a few species, and to the economics and social consequences of forest harvest. Jerry Franklin and Richard Forman (1987) have added a further dimension to the clear-cutting discussion by showing that the landscape pattern created by clear-cutting may be as important to forest survival and growth as the size of the cut patch, the location of roads to the patch, and other factors.

Their landscape analysis is based on a spatial model organized like a checkerboard (fig. 8.2). Progression of clear-cutting can be depicted by the location of patches of various sizes and shapes on the board. The authors assumed a thousand-square-hectare area, divided into ten-hectare cells, with a cell removed at random by each logging operation. As cutting progresses the landscape characteristics change, but these changes are not linear and continuous. For example, interpatch distance—the distance a fire or a seed must travel to move from one patch to another—is low until the 50 percent checkerboard point is reached. After this point the isolation of patches increases rapidly.

When the forest is between 30 and 50 percent cut-over, a threshold in the amount of edge is observed. Small patches have a large amount of edge or border where the environmental conditions are between those of an open area and a forest. Interior forest species may not be able to live in these edge habitats. As the amount of edge increases, the interior forest habitat decreases more rapidly than the total amount of forest area. Patches are also more susceptible to having trees blown down by wind, so that a highly fragmented forest might be blown over and all fragments be lost while large intact patches would resist wind. Thus it appears that the shape and distribution of landscape patches are important in natural resource management. Apparently there is an optimum patch size and patch distribution for each forest habitat that will yield the greatest goods and services on a sustained basis.

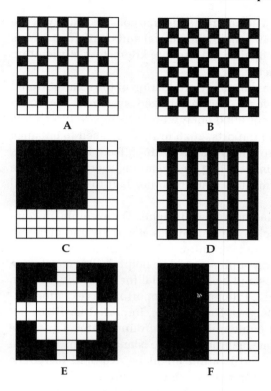

Fig. 8.2 Patterns of clear-cutting under different landscape models. *A* to *C* show a progression of clear-cutting using the dispersed-patch model, in which areas are selected for cutting so as to be regularly distributed. The points where 25, 50, and 75 percent are cut over are indicated. *D, E,* and *F* show patterns of cutting at the 50 percent point using a single-nucleus model, a four-nucleus model, and a parallel cutting system. Redrawn from Franklin and Forman 1987.

Implications

I have been discussing relatively small-scale landscapes in which the human impact came from outside. This approach has permitted me to focus on the theoretical development of the concept of landscape. It is misleading, however, because essentially all landscapes are affected by humans and the role of humans has to be treated as part of the landscape dynamics. There are many examples of such analyses. Here I will use a study of an island to illustrate the approach. Islands are valuable for many environmental studies, partly because they are well-bounded systems and one can monitor flows into and out of the system, and partly because they are usually relatively small. The island of interest is Gotland, which lies off the coast of Sweden about two hundred kilometers south of Stockholm (Jansson and Zucchetto 1978).

Gotland is about 130 kilometers long and 30 kilometers wide. It has had a long history of human occupation from 5000 B.C. Today it has a population of about fifty-four thousand people, twenty thousand of whom live in one town. Jansson and Zucchetto were especially interested in the energy dynamics of Gotland (fig. 8.3). The natural landscape consists mainly of the Baltic Sea coastal systems (2449 square kilometers) and coniferous forests (1260 square kilometers), in a total of 5517 square kilometers. Agriculture accounts for 871 square kilometers.

The energy studies compare the energy production of the solar-powered systems against that of the fossil-fuel subsidized systems. The solar-powered systems produce about 3000 energy units, compared to energy imports of 14,441 units. Of the imports, 2026 units were used to enhance solar-powered production through subsidies for fuel, feed, farm families, forestry, and fisheries. Approximately half the remaining energy went to the urban system, including the military, and the other half to the electrical industry.

The study of Gotland yielded many new insights. For example, in comparison with Sweden as whole, Gotland used less electricity and more oil, at lower efficiency in industry and agriculture. Clearly, energy import is a major cost and alternative energy sources could be helpful. This conclusion led to the suggestion that the island had a potential for using wind power, which had been used in the past. Problems and opportunities exist for coupling natural and cultural energies. For example, imported fertilizer improves agricultural yield but contaminates the ground water. Thus, one change produces a ripple of change throughout the island system. This conclusion suggests that organic fertilization might be explored to reduce contamination. Planning and anticipating change are always difficult, but models such as figure 8.3 are helpful in showing relationships between landscape elements.

Readings

Forman, Richard T., and Michel Godron. 1986. *Landscape Ecology*. New York, Wiley.

Hansen, Andrew J., and Francesco di Castri. 1992. *Landscape Boundaries: Consequences for Biotic Diversity and Ecological Flows*. New York, Springer.

Iverson, L. R. 1988. "Land-use Changes in Illinois, USA: The Influence of Landscape Attributes on Current and Historic Land Use." *Landscape Ecology* 2: 45–62.

Johnston, C. A., and R. J. Naiman. 1987. "Boundary Dynamics at the Aquatic-Terrestrial Interface: The Influence of Beaver and Geomorphology." *Landscape Ecology* 1: 47–58.

Zonnenveld, Isaak S. 1995. *Land Ecology*. Amsterdam, SPB Academic Publishing.

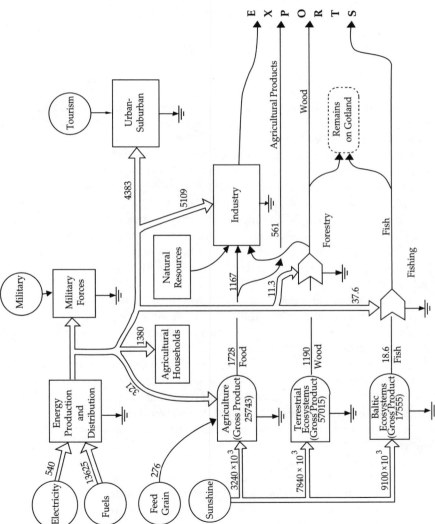

Fig. 8.3 Distribution of energy (in 10⁹ kilojoules) to the subsystems of the Gotland island, Sweden, ecosystem. The sources are shown as circles, production centers as bullets, and consuming systems as rectangles. The arrows show flows into and among subsystems and out to export. The flows are in energy units. Redrawn from Jansson and Zucchetto 1978.

9

The Watershed

In discussing the decomposition of the biome into landscapes, I used information about the dominant vegetation, soils, and topography to distinguish landscapes. But other criteria and other categories of landscape units might also be relevant. If, for example, our interest shifted from the biological features discussed in Chapter 8 to the chemistry of the landscape, we would be especially concerned about the flow of water and the sources of chemical elements in the lithosphere. This subject has been called environmental geochemistry or biogeochemistry (Fortesque 1980). The landscape appropriate to study of environmental geochemistry is the watershed. Landscape and watershed both refer to subdivisions of biomes. Landscapes can be subdivisions of large watersheds, and small watersheds can be part of landscapes.

Watersheds are an area of land over which water moves under the force of gravity from topographic highs to topographic lows (fig. 9.1). Usually water collects in streams and rivers and eventually enters the sea, where it deposits sediment and chemicals carried in the water. Watersheds are dynamic landscape units that emphasize water movement. They range in size from large river basins, such as the Mississippi River basin that occupies the center of the United States from the Appalachians to the Rocky Mountains, to small watersheds at the head of a stream.

It is relatively easy to determine the boundary of a watershed because the water that falls as precipitation flows across it through force of gravity. At the boundary of a watershed, falling water flows in one direction into the watershed or in another direction into a neighboring watershed. The location of a watershed boundary is usually unambiguous. Of course, boundaries are less clear on a flat surface, and in desert regions watersheds may

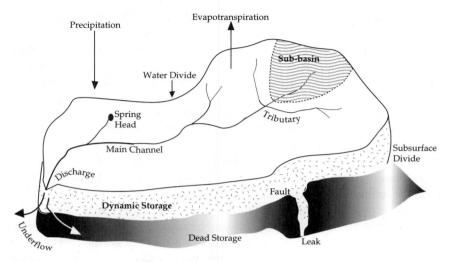

Fig. 9.1 Diagram of a typical watershed, showing parts and flows in the drainage basin.

be connected to interior salt lakes or flats that accumulate the products of weathering and leaching.

In the watershed, water acts on the ground surface, physically moving soil, rock, and organic matter as it flows across it. In addition, water dissolves chemicals from the rock and soil and carries them in solution from the land to streams or rivers. In this way, water physically and chemically sculpts the landscape.

The physical and chemical erosion of the earth surface is strongly influenced by the biota. Living and dead vegetation can control the rates of these processes. Presence of vegetation reduces erosion. Thus, the study of the dynamic processes on a watershed involves the integration of information from the sciences of ecology, geology, hydrology, and environmental chemistry.

Dynamics of Water

The first step in the study of the dynamic behavior of water on a watershed is to determine the input and output of water from the system. The principle underlying this operation is the physical law of the conservation of matter. Like the laws of thermodynamics, the law of conservation of matter is a fundamental control on processes operating on the Earth surface. The law of conservation of matter states that matter can be neither created or destroyed. It can change state and form, but the chemical elements making up matter always persist and can be collected and counted.

The consequence of this law is that the study of the dynamic processes on a watershed or other ecological systems always considers the balance of physical, chemical, or biological elements. The balance involves comparison of the input and outgo to and from the system. In the case of the watershed these include precipitation, evapotranspiration, discharge, and underground flows.

Rain and snow gauges are used to collect and measure input water to a watershed. This is a technically complicated task, because water does not fall uniformly over the ground surface, nor does it all fall as rain or snow. For example, fog condensation on leaves and branches of vegetation can be an important input in certain climates. Further, rainfall at the surface is extremely patchy and the rainfall amount per area is highly variable.

The output of water is divided into three principal flows. Some water is evaporated or transpired from the vegetation and soil and reenters the atmosphere as water vapor. Some water flows across the ground surface and directly enters the exit stream. And some water infiltrates the soil and rock and contributes to the groundwater. Water taking this latter route may reappear on the surface or in streams as springs or seeps downslope, or it may go deep into the basement rock and form underground water systems.

The surface water leaving the system is often measured at a small dam (or weir) located at the base of the watershed. If the rock under the watershed is solid and not cracked, it can prevent water from flowing into the deeper ground strata. This creates an ideal situation for study because the dam can be sealed to the basement rock, so that all the water that does not leave through atmospheric pathways will appear at the dam, where it can be measured. Evaporation and transpiration into the atmosphere from leaf and soil surfaces can be measured directly with suitable instruments. If we have data on water flow on two of the three routes, the flow on the third can be calculated by subtracting the known from the total flow. We can use these relationships to test our calculations. For example, if a hydrologist measures water flow across a weir and in evapotranspiration and finds that the quantity moving on these two routes equals the input from precipitation, then this is good evidence that the watershed is closed and water is not leaking into deeper rock strata.

Studies using this approach have shown that the proportionate flows of water over these pathways are relatively constant over time. For example, at Hubbard Brook Experimental Forest in New Hampshire, ecologists (Likens et al. 1977) measured water flows from 1956 to 1974 and found considerable variation in precipitation and stream flow from year to year (fig. 9.2). For example, 1964–1965 was a dry year, while 1973–1974 was quite wet. There was no particular pattern in precipitation from year to year. Rather, the variation in the data reflected larger-scale processes in the biosphere

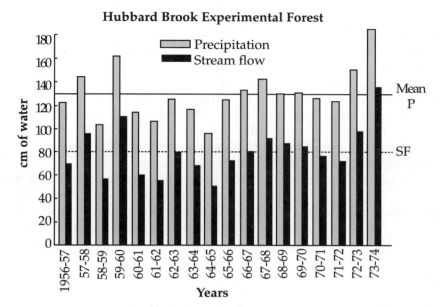

Fig. 9.2 Annual precipitation and stream flow for an undisturbed forested watershed on Hubbard Brook Experimental Forest, New Hampshire. From Likens et al. 1977.

that control climate. Despite these variations, however, the processes of input and output were tightly coupled. When one increased, so did the other. The difference between them (evapotranspiration) stayed about the same each year. The proportion of water leaving the watershed on output pathways did not change greatly because vegetation requires a fixed amount of water in transpiration to survive. This amount is used regardless of the difference in input (although individual species of plants do have mechanisms to regulate water loss). If the input falls below the requirement, the plants exhibit drought stress and eventually die. The difference between evapotranspiration and water input is stream flow.

Streams and rivers in the watershed and river basin have their own special ecological features. Probably of most significance is the fact that they form a continuous linked system from headwater to the ocean or lake at their mouth (Vannote et al. 1980). The sections of this continuum have unique characteristics. For example, the headwaters in a mountainous terrain are fast-flowing over rocky substrate, with overarching trees and shrubs along the banks, abundant leaf litter in the stream channel, and many organisms, such as crayfish, that shred the organic matter into smaller particles. In the lower reaches of a river water flow is slower, the temperature is higher, there is much less input of organic matter, and the water often carries more sediment. The organisms shift to those able to use finer organic matter particles in the water. Finally, at the lowest portions of rivers, which may be so large as to appear to be lakelike, the water flow is slower, the water is warmer and more turbid, and the organisms depend upon organic production within the river itself or from flood plains bordering the river. Indeed, in very large rivers such as the Amazon, the flood plains are essential elements of the river system, providing nutrients, organisms, and food for the river life.

Since each section of a river has its own fauna and flora, the nutrients, which are continuously moving downstream, also may be taken up and recycled within the section. This combination of downstream flow and recycling produces a spiral path for many elements as they move from the headwater to the ocean. All of these features will be dominated by the occasional flood event, which not only inundates the floodplains and uplands bordering the river but also moves large quantities of sediment, nutrients, and organisms downstream. Life in these systems is strongly influenced by physical events within and outside of the watershed.

Watershed studies like those at Hubbard Brook have been carried out in many different biomes. They are exceptionally valuable because they provide data on discrete systems, the watersheds, which are large enough to be significant from a practical standpoint and yet small enough to be used experimentally. The latter function has allowed ecosystem studies to become an experimental science. As an example, the vegetation in a watershed can be altered and the role of that vegetation on water flow determined. Removing trees and replacing them with grass reduces the flow rate over the evapotranspiration route and increases flow to streams. Some communities with limited water supplies have used this observation to enhance water flow to water storage facilities and provide a more dependable supply of drinking water—for instance, by changing the watershed vegetation from forest to grassland to increase stream yield.

Biogeochemistry

The waters flowing into and out of the watershed carry a variety of chemical elements and are part of the elemental cycles that link the water, land, oceans, and atmosphere into a single interacting biosphere. The interaction of this water with the lithosphere is an active chemical process, called weathering. Through weathering the surface of the lithosphere is transformed into soil. Soil is a material created through the interaction of life, environment, and substrate. It represents a gradient of properties with depth. At the surface it contains a large organic component and is strongly influenced by the biota. At depth it connects to the untransformed base-

Table 9.1 Annual mean concentration of dissolved substances in bulk precipitation and stream water in undisturbed forested watersheds on the Hubbard Brook Experimental Forest in New Hampshire from 1963 to 1974. From Likens et al. 1977.

Substance	Precipitation (mg/l)	Stream Water (mg/l)
H+	0.073	0.012
NH$_4$+	0.22	0.04
Ca++	0.16	1.65
Na+	0.12	0.87
Mg++	0.04	0.38
K+	0.07	0.23
Al+++	——	0.24
SO$_4$-	2.9	6.3
NO$_3$-	1.47	2.01
Cl-	0.47	0.55
PO$_4$-	0.008	0.0023
HCO$_3$-	0.006	0.92
Dissolved silica	——	4.5
Dissolved carbon	2.4	1.0
Total	**8.64**	**18.70**

ment rock. The resulting soil profile may be tens of meters deep in the tropical rain forest, where the weathering processes have been active for long periods, or it may be a few centimeters deep in a dry forest.

The rate of weathering depends on many factors, including the chemical composition of the input water to the watershed. For example, at Hubbard Brook the precipitation was relatively rich in sulfate and nitrate, which is derived partly from industrial pollution (table 9.1). These elements combine with water to form acids that can enhance weathering. The acids in turn combine with the basic elements in soil and rock and make them susceptible to chemical leaching. Sodium is a chemical element that is readily leached from rocks. Calcium, magnesium, and potassium are more slowly leached, while iron, aluminum, and silicon are relatively resistant to leaching. The type of soil and its fertility depend on the rate and extent of weathering.

When calcium is leached from the soil, it is carried by water into an exit stream where it becomes available to aquatic life or is carried to the ocean and deposited in ocean sediments. The stream water leaving the Hubbard Brook watershed was rich in calcium but also contained a variety of other elements. The loss of chemical elements from the watershed exceeded the input from precipitation and represents the weathering or decomposition of the lithosphere. If these processes were to continue over sufficient time, the surface of the lithosphere theoretically would become almost flat. This theoretical endpoint is called a peneplain. It is seldom if ever achieved in fact.

In my discussion of the landscape I treated the biota as a structure, called a life form, and used it to identify biomes. At this finer scale of the watershed the biota plays a different role and must be described in more detail. The vegetation not only controls the rates of water movement in evaporation and transpiration but also influences the rate of weathering of the lithosphere. This role is both positive and negative. In a positive sense, the vegetation controls the surface flow of water, reducing erosion of the soil surface. The dead plant material forms a surface protecting the soil from direct impact of rain and forms dams across the slope or in streams which slow the flow of water and create pools. In a negative sense, the vegetation produces organic acids that increase the leaching rates of soil chemicals and the plant roots physically break the surface rock and enhance erosion.

These processes depend a great deal on the kind of plants, animals, and microorganisms growing on the watershed. Experiments at the USFS Coweeta Hydrologic Laboratory in Franklin, North Carolina (Swank and Crossley 1988) have shown that the type of vegetation cover on a watershed strongly influences the rates of loss of the chemical elements. One watershed growing a forest of hardwood deciduous species was cut and planted with white pine, another watershed was cut and recut at regular intervals (called coppicing), and a third was clear-cut and left to regrow into a forest. Comparison of each of these watersheds against a hardwood-forested watershed, which served as the control, showed the net gain or loss of ions (table 9.2). All of the treated watersheds lost nitrogen, as nitrate and ammonia, and did not differ from the control for phosphate. For the other ions, the stands behaved differently. The clear-cut stand, allowed to revegetate naturally, lost ions at relatively high rates. This effect was due partly to the fact that the treated stands were growing rapidly and incorporating these chemicals in their wood. There was less woody growth on the grass-to-forest succession watershed.

The coupling of forest and stream can be very tight. At Coweeta the ecologists were alerted to an incidence of defoliation of the trees by insects through a change in the nitrate level in the stream water (fig. 9.3). On one watershed (shown in the upper graph), insect consumption of leaves was unusually high in 1974. On another watershed (the lower graph), defoliation events were observed repeatedly. This high rate of consumption was correlated with a high rate of nitrogen loss. Insects eat leaves and defecate frass, which falls to the forest floor. This material is a substrate for bacteria

Table 9.2   The net gain or loss of ions on experimental watersheds at the U.S. Forest Service's Coweeta Hydrologic Laboratory, North Carolina. Values are in kilograms per hectare per year. From Swank and Crossley 1988.

Vegetation Type	NO_3-N	NH_4-N	PO_4	Cl	K	Na	Ca	Mg	SO_4
White pine	−0.7	−0.1	0.0	+1.8	+1.7	+4.4	+2.3	+1.2	+0.7
Hardwood coppice	−2.2	−0.3	0.0	+1.0	+0.9	+2.8	+0.2	−0.6	+0.8
Grass-to-forest succession	−6.8	−0.1	0.0	−4.4	−0.7	−0.1	−3.3	−2.6	−1.0

and other microorganisms that mobilize nitrogen, which then becomes available for transfer out of the watershed in stream water.

Nitrogen has been studied on landscape-sized agricultural watersheds, because it is an important fertilizer for crops, and leaching of unused fertilizer into the groundwater or into streams may cause health hazards. Byron Kesner and Vernon Meentemeyer (1989) studied the flows of nitrogen across a landscape in south Georgia, near Tifton. The study area on the Little River Watershed was 114 square kilometers in size. A large amount of nitrogen is used as fertilizer in this important farming region (fig. 94). The pools of nitrogen in the atmosphere, vegetation, and soil are large relative to the fluxes. But note the difference between the flows into and out of the watershed in streamwater. Very little nitrogen is lost to the stream. Most of the nitrogen leaves the watershed in the harvested crop.

An interesting finding in this study concerns this low loss of nitrogen to the streams. In this landscape the streams are bordered by swamp forest. As nitrogen flows from upland fields to the stream, it is intercepted by the forest. The trees take up nitrogen in their growth and store it. Further, the low-lying lands along the streams flood in rainy periods. During floods the streamwater covers the forest floor, cutting off the oxygen to the litter organisms and bacteria. These conditions are suitable for denitrifying bacteria, which utilize the nitrogen in their metabolism and convert it into its gaseous state. This nitrogen gas is lost to the atmosphere. These two processes use so much nitrogen that very little is left to pollute the stream.

Implications

Water is required for life. The ordinary person needs about 2.5 liters of drinking water a day. We can do without food for several weeks but will die without water after a few days. In addition, water is required for all phases of life activities. The average American uses more than 300 liters of water per day, in addition to drinking water. Further water is required for food

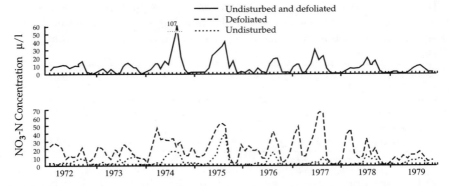

Fig. 9.3   The mean monthly concentration of NO_3-N in streamwater of three watersheds at Coweeta Hydrological Laboratory, North Carolina. The upper graph shows a watershed that experienced insect defoliation in 1974. Note the rapid and unusual increase in stream nitrogen associated with this event. The lower graph describes the pattern on a watershed that did not experience defoliation. Data from Swank and Crossley 1988.

production and for industry. Sandra Postel and her colleagues (1996) estimate that humans now use 26 percent of the total terrestrial water flow through plants (evapotranspiration) and 54 percent of the accessible runoff in rivers. If a material such as water is required for life, why would we allow it to be wasted, polluted, and otherwise misused? Why would our rivers and lakes be allowed to become waste dumps and be used to transport sewage? One would think that water would be highly valued by humans, yet the evidence suggests the contrary. How do we calculate the value of a resource such as water? How do we manage watersheds to protect our water supply?

Value implies a valuer; however, value also resides in that which is valued. Value may be based on the fact that the object or process provides goods or a service to the valuer. This type of value is called instrumental value. Water has instrumental value to humans in the context developed above. It is essential to our life. Our bodies are composed mainly of water, and water transports nutrients and wastes through our veins. Water also transports wastes, supports our ships, and has other values to human society. These values can be expressed in economic terms.

For instrumental value to be recognized, an individual or group of individuals must have responsibility for the object or process. Responsibility is often expressed through property rights. The individual with such responsibility charges a fee or equivalent in exchange for use of the resource. If no individual owns or is responsible for a resource, it is likely to be misused or destroyed. This has been the fate of rivers, lakes, and water generally. The

Nitrogen Mass Balance
Little River Watershed-F
Turner County, Georgia, 1981

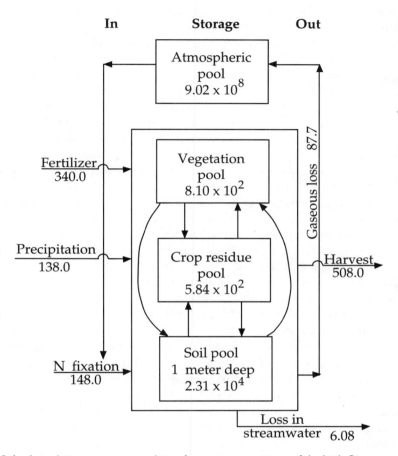

Fig. 9.4 Annual nitrogen storage and transfer among components of the Little River Watershed, Tifton, Georgia. The values are in metric tons of nitrogen. Data from Kesner and Meentemeyer 1989.

concept of the common good is little respected in a culture that adulates individualism.

The objects and processes of nature also have another kind of value, which resides in them without reference to human need or interest. Intrinsic value is derived from the inherent role or character of the object or process. If it is a biological object, intrinsic value derives from its evolutionary history, its natural selection, its place within ecological communities, or its integrity as an ecosystem (Westra 1994). If it is a physical object or process such as a mountain or a river, its value is derived from the geological processes of uplift and erosion and from the hydrologic cycle.

Intrinsic value does not derive solely from the existence of an object or process. It also represents the value of being part of natural systems and playing a role in natural processes. Ultimately, intrinsic value derives from the linkages and mutual dependencies the object or process has with other objects in ecological systems. Clearly, an object or process could have both instrumental and intrinsic value. However, many features of the natural world have intrinsic value but lack obvious instrumental value to humans.

Barbara Tuchman (1984) defined folly as taking a course contrary to what one knows is in one's best interest. It is folly, in Tuchman's sense, for humans to ignore intrinsic values. The challenge for us is to move from a strategy of making decisions based entirely on instrumental values toward respect for intrinsic values that may have no direct connection with humans but exist in ecosystems. Because we are connected directly and indirectly to these systems via natural links and connections, ultimately our respect for intrinsic values benefits us and they become instrumental values.

The misuse of water stems from many sources: the assumption that it is an unlimited resource, a failure to appreciate water's importance and the roles it plays, the lack of an understanding of the linkages of water and land, and the fact that water is a commonly held rather than privately owned resource in much of the world. Water is among the most abused resources we have, and in many places its scarcity limits human well-being. An understanding of the concept of the watershed and water flow through the hydrologic cycle helps in understanding that water connects the most distant systems across biomes.

Readings

Bolling, David M. 1994. *How to Save a River: A Handbook for Citizen Action.* Covelo, Calif., Island Press.

Clarke, Robin. 1993. *Water: The International Crisis.* Cambridge, MIT Press.

Feldman, David Lewis. 1991. *Water Resources Management: In Search of an Environmental Ethic.* Baltimore, Johns Hopkins University Press.

Fortesque, John A. C. 1980. *Environmental Geochemistry: A Holistic Approach*. New York, Springer.

Gleick, Peter H. 1993. *Water in Crisis: A Guide to the World's Fresh-Water Resources*. New York, Oxford University Press.

Myers, Mary Fran, and Gilbert F. White. 1993. "The Challenge of the Mississippi Flood." *Environment* 35, no. 10: 6–36.

Naiman, Robert J. 1992. *Watershed Management: Balancing Sustainability and Environmental Change*. New York, Springer.

Watson, Lyall. 1988. *The Water Planet*. New York, Crown.

White, I. D., D. N. Mottershead, and S. J. Harrison. 1984. *Environmental Systems: An Introductory Text*. London, Allen and Unwin.

Worster, Donald 1985. *Rivers of Empire: Water, Aridity and the Growth of the American West*. New York, Pantheon.

10

The Ecotope

At the smallest level of spatial scale are those ecosystems in which we live, work, and play. They occupy from tens to thousands of square meters and may consist of a lawn, a patch of forest among agricultural fields, a pond, or the south slope of the highest watershed on a mountain stream. Because these systems are small, we can interact intensely with them, observe them in great detail, dominate and convert them to our purpose. We call these fine-scale systems ecotopes.

An ecotope is an ecosystem with homogeneous properties of interest to the observer. If we were interested in forest vegetation, a patch of ground with trees throughout would represent a forest ecotope. If we were concerned about the flow of water over a surface, a south-facing hill with the same slope, aspect, and vegetation would be an ecotope. An agricultural field that had been plowed and leveled and planted to a single crop would also be an ecotope. With the ecotope, we have reached a point in the spatial hierarchy where no further decomposition of ecological systems is possible if we follow the principle of self-similarity. The ecotope is the smallest unit of the spatial hierarchy.

Ecotope Structure

The ecotope is small enough that almost every property owner controls one ecotope and some control several or many ecotopes. We are highly conscious of the flora and fauna of ecotopes and are frequently experts on the quality of their soils, their productivity, and their capacity to receive and hold water. We fertilize them, burn them, plow them, and manipulate them in countless ways. They include our favorite places, as well as the trash heaps where we dispose of refuse and junk. Because we live in and use ecotopes daily, we have an intuitive sense of how they are built and work. The scientific analysis of ecotopes goes beyond this intuition and focuses on aspects of structure, function, and capacity to change. I will examine these features of ecotopes in this and the following chapters.

We can describe system structure from several points of view. First, we can view the ecosystem from above, as in an aerial photograph, or from the side in profile (figs. 10.1 and 10.2). I considered the aerial perspective in Chapter 8, noting that the landscape appeared as patches and corridors of vegetation on a background matrix consisting of the dominant form of land use. The side perspective reveals that the system is composed of strata. A grassland may be composed of only one or two strata of plants, while a tropical rain forest may have many strata of trees. These strata can be described in terms of the amount of area occupied by them, the weight of the components in that stratum, or its species composition. Strata include the parts living in the soil, as well as those in the air or water. The different strata create niches available to other species so that they enhance the number of species living in the system. The life-form concept that I used to characterize biomes represents structure in this sense.

Second, we can identify and list the species that are present in the ecotope. A count of the species is a measure of their diversity. Ordinarily, these counts are made on a replicated series of quadrat or transect samples. The relationship between the number of new species of organisms encountered in the samples as the area sampled increases is called a species-area curve (fig. 10.3). This curve usually increases to a point and then remains constant, indicating that all of the species likely to be present have been encountered. Usually the samples will have missed some of the rare species, but on average the number observed at the breakpoint in the curve is a reasonable measure of the diversity of the organisms in the ecotope. Not all species-area curves become constant: rain forests often have curves that continue to increase.

The measure of diversity can be improved by determining the abundance of each species on the list. In this way the dominant or most abundant species can be distinguished from rare species. The simplest measure of diversity is the number of species divided by the number of individuals. This measure is useful for many purposes, but it can be misleading if the species and individuals in several ecotopes are arranged differently. For example, two ecotopes might have ten species with a hundred individuals. In one ecotope each species may be represented by ten individuals, while in the other ecotope one species may have ninety-one individuals and nine species may have one individual. The species per individual index would be the same for both cases but the actual relationships would be exceedingly different. Species-diversity indices based on mathematical theory correct for this and other bias (Peet 1974).

Fig. 10.1   Aerial view of a tropical moist forest in the Republic of Panama. Photo by Peter McGrath for Golley et al. 1975.

Fig. 10.2 Cross-section drawing of a tropical moist forest, showing the multiple strata of the canopy. Three strata are pictured: an emergent tree above a continuous forest canopy and a lower canopy near the ground. Drawn by Priscilla Golley for Golley et al. 1975.

Third, the components of the ecosystem can be treated as receptacles of energy or chemicals. In this case it is necessary to weigh individuals of each species and collect environmental samples and then determine their energy or chemical content in the laboratory. The weight or physical mass of organisms is called biomass. Biomass includes both living and dead tissues. It represents the chemical elements and energy in a biologically controlled form where the physical environment cannot act on it directly. The quantity of biomass in an ecotope may be large. For example, in a tropical forest the biomass may exceed eleven hundred metric tons per hectare. The physical chemical material that is located in the nonbiomass part of the system also may vary in amount and can be part of the inventory.

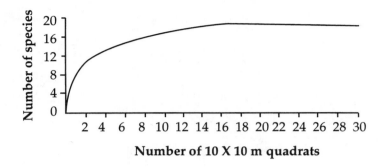

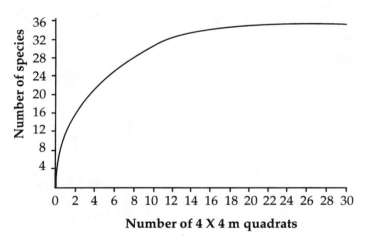

Fig. 10.3 Species-area curves for oak-hickory forests in North Carolina. The upper curve represents trees sampled on 10 × 10 meter plots. The lower curve represents shrubs sampled on 4 × 4 meter plots. Redrawn from Oosting 1948.

    A full structural analysis will measure several of these parameters together. For example, it may be of interest to know how the biomass is distributed in space and time. Since light is a limiting factor for plant growth, the horizontal distribution of biomass is influenced by the availability of light. Most of the leaf biomass in a forest canopy is near the source of light (fig. 10.4). Plants compete for the light resource and build a large stem and root biomass to support the leaves and branches. There is a tendency for trees to grow taller and taller, but there is a limit to the height they can grow. Tree height is partly controlled by the capacity of evaporative energy at the leaf surface to pull water up the stem from the soil and partly by the construction of forms of wood that can withstand strong winds. In the lat-

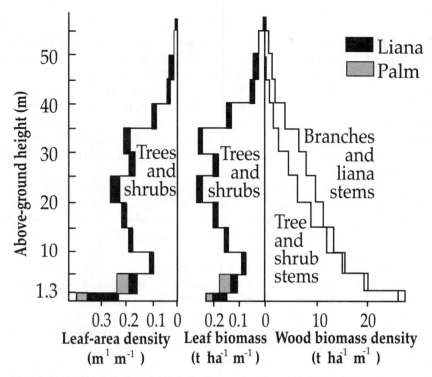

Fig. 10.4 Profiles of leaf-area density, leaf biomass, and wood biomass density in a tropical lowland rain forest at Pasoh, peninsular Malaysia, based on clear-felling a plot 20 × 20 meters. From Kira and Yoda 1989.

ter case, the construction material of the plant cell is cellulose and lignin, which are very strong materials. Further, the plant cells may be twisted in complex spirals which provide a tension in the stem that permits considerable bending and twisting under wind pressure but resists breakage. Trees may reach a hundred meters or more in height.

The chemical characteristics of the organisms are derived ultimately from the soil or water in which they live. Therefore, a structural analysis usually involves study of the soil or substrate. Examination of soil chemistry and measurement of the amount of total chemical and available chemical elements is a difficult and technically complex subject (Fortesque 1980).

The pattern of distribution of planktonic organisms in the ocean parallels the pattern of leaves in a forest. The distribution curve of biomass with depth is similar. However, in the case of the hydrosphere the pattern is caused by the extinction of light as it passes through the water column. Plankton float near the surface, where the light energy is most abundant.

Abstract Structural Descriptions

The number of species that may be encountered in a typical ecotope is probably many thousands—we don't have a full listing of species in common systems. It is very difficult for an ecologist or even a group of ecologists to examine each species individually and then sum the individuals and the species to gain an understanding of the structure of a whole system. Instead, ecologists have developed a variety of abstract ways to represent the biotic structure without actually counting and studying the biology of each species. These measures can be used to characterize functional features of the system.

One way to deal with the problem of a large number of poorly known species is to base the structural analysis on functional properties of the ecotope. For example, a central process of all living systems is feeding. Feeding links the species together and represents the passage of energy and nutrients from one species to another. A British ecologist, Charles Elton, visited the arctic island of Spitsbergen in the 1920s and studied the feeding patterns of the relatively few species that could survive in the harsh environment. Elton (1927) arranged these species into groups, called trophic levels, based on their distance from solar energy as the primary source of energy for the system. He then compared the biomass or abundance of organisms in each level. He found that the levels formed a pyramid, with a wide base of plants, fed upon by herbivores, which constituted a smaller biomass than the plants. The herbivores supported carnivores, which had a smaller biomass than the herbivores. The carnivores supported a few top carnivores.

Elton's pyramid of biomass has been found to be representative of many terrestrial ecosystems (fig. 10.5). However, in aquatic systems the floating photosynthetic organisms do not require support and their biomass may be quite small. They are fed upon by slightly larger animals, which support yet larger fish. Thus, the pyramid of biomass may be reversed in some aquatic systems.

The pyramid-of-biomass concept does not deal adequately with bacteria and other small organisms. Yet bacteria may be very important because they reduce organic matter to forms in which its chemical constituents can be reused by the biota (Pomeroy 1974). In a sense, the entire above-ground pyramid of biomass that represents the elaborate structure we see and enjoy is paralleled by an equally elaborate structure devoted to deconstruction of the above-ground material. However, this structure is characterized by small organisms that mainly operate chemically on the organic material.

Ecologists have used other abstractions to describe the structural linkage of species. For example, the feeding process can be expressed as a

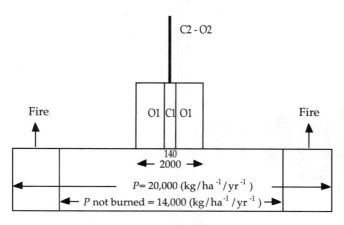

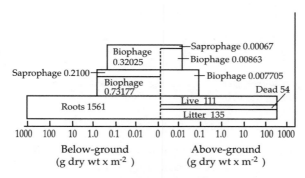

Fig. 10.5 Examples of ecological pyramids. Above: a pyramid of energy content in the biomass of the Lamto savanna ecosystem, Ivory Coast, Africa. *P* is primary production, including the amount of plant material burned in fires. *C1* is the energy in primary consumers, *O1* is that in decomposers of plant material. *C2* and *O2* are consumers and decomposers of animal material. Data from Lamotte 1975. Below: a pyramid of biomass in the North American grassland showing above- and below-ground organisms. Biophage means consumers of living tissues and saprophage means consumers of dead tissues. Data from French et al. 1979.

linked set of food transfers or a food chain. Food chains are usually only four or five links long (fig. 10.6).

Ecologists are still arguing about why this is so (Pimm 1982). One explains that energy is lost to heat at each step (recall the second law of thermodynamics, which results in the loss of as much as 50 to 90 percent of the energy for further work at each step); therefore, relatively little energy would be left to support the next species in the chain. After three to five transfers all the energy would be used up. Another ecologist suggests that short chains are more stable and that long chains, if destroyed, would re-

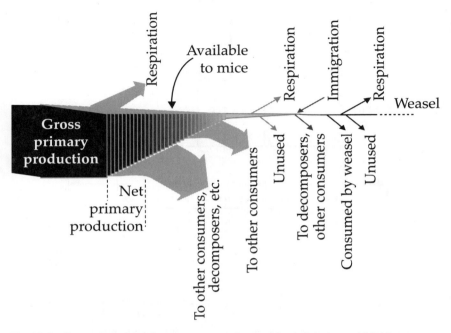

Fig. 10.6 Energy flow through a grass, mice, and weasel food chain in an old-field community. The relative sizes of the arrows indicate quantity of energy flowing through each link. Based on Golley 1960.

quire the invasion of new species to maintain the integrity of the chain. Thus, chains of three to five links have survival value and persist. Probably there are a variety of reasons for food chain length.

Food chains have been criticized by ecologists because there are few linear-flow sequences where one species feeds exclusively on the next. Usually a species is fed upon by a variety of herbivores or carnivores. This means that the feeding relationships of ecosystems are better described as food webs or networks of food chains than as linear sequences of feeding. Charles Elton was one of the first to describe the transfer relationships of an ecosystem as a web (fig. 10.7). He and V. S. Summerhays expressed their food web in terms of nitrogen, a key element for system growth and development.

Implications

Because ecotopes are small in spatial dimensions and humans can dominate or create and manage them, we tend to focus on the biological patterns of the flora and fauna more than on the physical-chemical environmental

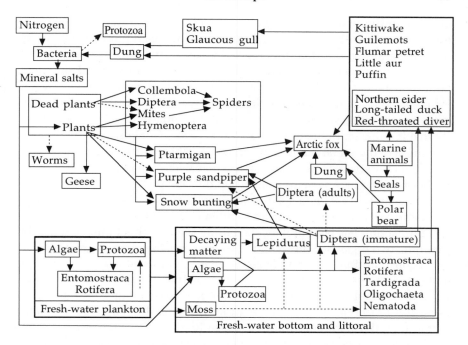

Fig. 10.7 An example of a food web, described by Summerhays and Elton 1923, from Spitsbergen Island. The diagram reports flow of nitrogen between compartments.

patterns. At this level of scale the abiotic patterns are defined as relatively uniform, whereas at the level of the biome, landscape, and watershed, the environment plays a dominate role in definition of the system. At this fine level of scale the biological features of the system emerge as most visible and significant.

Because the plants and animals in the ecotope interact, they are frequently referred to as a biotic community. The word *community* emphasizes the interactions between the flora and fauna, represented by the profile diagram of strata or the abstract food web showing feeding interactions. W. C. Allee and his colleagues, in their classic *Principles of Animal Ecology* (1949), stated: "In large, the major community may be defined as a natural assemblage of organisms which, together with its habitat, has reached a survival level such that it is relatively independent of adjacent assemblages of equal rank; to this extent, given radiant energy, it is self-sustaining." The community concept has played and continues to play a major role in both theoretical and applied ecology (Noy-Meir and van der Maarel 1987, Strong 1983).

The concept of community implies that individual organisms interact. Direct interaction involves face-to-face or side-by-side contact between two or more individuals. In this circumstance individuals may help one another, compete with one another, feed on one another, reproduce with or ignore one another. Information on species diversity helps us predict what the neighbors may be like, assuming that the more dominant and abundant are more likely to appear next door, but chance plays a role, as always. When we add up all the interactions, we express them in a food web diagram or something similar, which loses all the intense life-and-death activity that is going on in the face-to-face encounters. Our abstractions are a bit like those of the social sciences when the horrible reality of violence is represented by statistics telling us that crime is declining in our town.

Bernard Patten, a systems ecologist at the University of Georgia, has created a metaphor that helps us link an abstract view of the community with the complexity of real interactions. Patten (1985) points out that every direct interaction has behind it an expanding set of indirect interactions. For example, two organisms meet and one consumes the other. Because of that act of consumption, one is no longer available for reproduction and there is a potential numerical impact beyond the feeding impact. Further, the lack of an individual capable of reproduction may give another individual an opportunity to reproduce, allowing that individual to contribute genetic material to the population and increase its diversity, and so on. Patten points out that direct interaction generates a cascade of impacts much like the spreading circles in water when a stone is thrown into a pond.

I think that Patten's metaphor is extremely helpful because it creates an opportunity to reflect upon our own actions. We too are members of communities of organisms, as well as communities of humans. Our direct action may be as meaningless or thoughtless as pulling a leaf off a shrub during idle conversation. Yet our actions result in a cascading series of reactions.

Readings

Kira, T., and K. Yoda. 1989. "Vertical Stratification in Microclimate," pp. 55–71. In H. Lieth and M. J. A. Werger, eds., *Tropical Rain Forest Ecosystems*. Amsterdam, Elsevier.

Lamotte, M. 1975. "The Structure and Function of a Tropical Savannah Ecosystem," pp. 179–222. In F. B. Golley and E. Medina, eds., *Tropical Ecological Systems: Trends in Terrestrial and Aquatic Research*. New York, Springer.

Noy-Meir, Imanuel, and Eddy van der Maarel. 1987. "Relations Between Community Theory and Community Analysis in Vegetation Science: Some Historical Perspectives." *Vegetation* 69: 5–15.

Pimm, Stuart L. 1982. *Food Webs*. London, Chapman and Hall.

Pomeroy, Lawrence R. 1974. "The Ocean's Food Web: A Changing Paradigm." *Bioscience* 24, no. 9: 499–504.

11

Species Diversity

One of the ways we can describe the structure of an ecosystem is by listing the species that are present in it. If we compare two ecosystems and one has a short list of species and the other a long list, we might conclude that the second is more diverse than the first. However, the abundance of each species is an important datum missing from the comparison. A community with a few very abundant species and many rare species is fundamentally different from a community with many moderately abundant species. Thus, species-diversity measures include information about both the taxonomic variety of the community and the abundance of the species composing that variety.

Although the diversity of the biota has increased over geological time, the rate of increase has varied (Signor 1985). Periods of rapid evolution of new forms have alternated with periods of stasis and decline or extinction, such as occurred at the end of the age of reptiles in the late Cretaceous period, 65 million years ago (fig. 11.1). Peter Sheehan, of the Milwaukee Public Museum, divides the last 460 million years of living organisms into six periods, each lasting from 35 to 142 million years. Each period is terminated by a mass extinction event, followed by a recovery lasting from 3 to 8 million years (R. A. K. 1994). Some groups of organisms have been highly conservative and are still abundant on the Earth. Others have increased, been dominant, and then died out or exist as remnants of a once significant fauna or flora. Of course, we do not know the full history of the biota, because some species did not become fossilized and are not available for study. Our speculations about changes in diversity are based on the groups we can observe in the fossil record. What is worse, we do not even know the number of species on the Earth today. Norman Myers, the English conservationist, calls it a scientific scandal that we can measure the distance

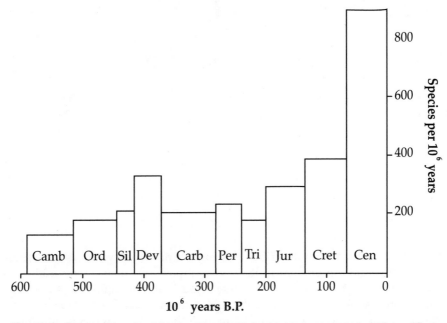

Fig. 11.1 Estimated species richness of fossilizable invertebrates over geological time. The names of the geological periods are abbreviated for each bar. The species richness is normalized to period length. From Signor 1985.

from the Earth to the moon with an accuracy of centimeters but we don't know the names of the Earth's biota, much less understand their biology.

Estimates of the diversity of the biota vary greatly. According to Myers (1979), we have identified only 1.7 million species. It is likely that there are at least 10 million more unnamed species. Recent discoveries in the tropics indicate that we have not understood how varied and numerous are the insects living in tree canopies. Considering these discoveries, some entomologists have estimated that there may be as many as 30 million species on the planet.

The largest proportion of the biota consist of insects (fig. 11.2), of which about two-fifths are beetles *(Coleoptera).* Vertebrates make up a small proportion of the total.

Species Diversity

Species diversity is the variety and abundance of the species living in an ecosystem and making up its biotic community. The species diversity at any site consists of those species from the potentially available species in

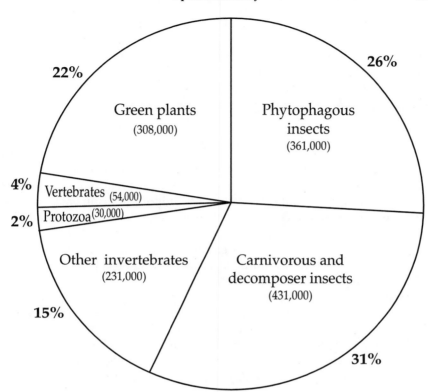

Fig. 11.2 The approximate number of known species on the Earth. The total number of species is unknown. From Southwood 1978.

the species pool of the biome and landscape which can migrate to the site, become established, and then reproduce themselves and maintain their population. Chance plays a role in which species arrive and survive at any given site. It is difficult for a species to enter an already established and thriving community, but where a community has been disturbed in some way a new species may enter and be able to establish itself. Further, some species respond to the special conditions of the habitat and evolve over time. Thus, new species and change in diversity may come from within or from outside the community. The species diversity of a community is a dynamic property and changes over space and time.

We have relatively few complete censuses of the biota of terrestrial and aquatic ecosystems. Least is known about the diversity of microorganisms (Schulze and Mooney 1993). Probably the vertebrate species and the higher plants are best known. For example, in a deciduous forest stand in the eastern United States there may be about a thousand species of plants and animals. This local diversity, however, has to be viewed against a larger regional diversity. In a stand, we might encounter twenty to thirty tree and shrub species out of a potential landscape species pool of one hundred and a regional species pool of three hundred. This means that if we examine a second stand, we are likely to encounter some new species.

Measuring Diversity

The number of species in an ecosystem is determined by sampling the fauna or flora in a series of representative samples, identifying the species in the samples, and counting the numbers of individuals in each species. Each sample is represented by a list of species and the numbers of individuals of each species. The number of species and their abundances depend on the size and area of the ecosystem, the shape of the sampling device, and many other factors. Sampling is a specialized subject and anyone desiring to learn more about it might start with Southwood (1978).

If we calculate the frequency distribution of species and individuals, we will observe that there are usually a few abundant species and many less abundant species. This frequency distribution is familiar in ecology. It means that a few species are dominant, because they find the conditions in the ecosystem especially suitable for their survival and growth, while most species are less abundant, because they are limited by inadequate resources, competition with other species, or a lack of vigor. All ecosystems are limited in their capacity to support species and individuals, and this limitation is reflected in the steepness of the species-abundance frequency distribution. Poor habitats have steep distributions, while fertile habitats have less steep distributions.

The Significance of Species Diversity

Species-diversity indices are useful indicators of the stresses to which the ecosystem is subjected. For example, a harsh environment tends to reduce the maximum diversity expected under optimum conditions. In a polluted river the characteristic species may be reduced or eliminated and the few species able to live in polluted conditions may become common.

Species diversity also is an indicator of the biological richness of ecological systems. In general, the highest diversity of higher plants, insects, and vertebrates are found in tropical rain forests. E. O. Wilson and F. M. Peter (1988) point out that tropical rain forests cover only 7 percent of the Earth's surface but contain more than half of the entire world biota. Under tropical climatic conditions a variety of life forms can exist, with multiple layers making up the tree canopy, and lianas and epiphytes being abundant (see Chapter 10). The more abundant the plant species, the larger the numbers of animal species and microorganisms that are supported. The food webs tend to be more complex in these environments. Moreover, the tropics probably did not experience the severe disturbance of Pleistocene glaciation. During the Pleistocene the tropics experienced climate change but species were not eliminated by advancing ice. These characteristics of the tropical environment produce conditions that support higher species diversity.

David J. Currie and Viviane Paquin (1987) examined the environmental factors controlling species richness of trees in North America. They reported that tree richness was highest in the Southeast and declined to the north and west. The maximum richness was 180 species in a quadrat 2.5 square degrees of longitude and latitude. The distribution of species richness suggests that climate must play a role in the pattern, and Currie and Paquin found that species richness was most highly correlated with realized annual evapotranspiration (that is, with the amount of water evaporated from the soil and transpired by the plants over a year). Evapotranspiration explained 76 percent of the variation in species richness. When they added other climatic factors and topography to the equation, they were able to explain 86 percent of the variation. Interestingly, historical factors such as glaciation and dispersal were not important. Why is evapotranspiration such an important variable? Currie and Paquin suggest that evapotranspiration is highly correlated with terrestrial plant production and that production is a measure of the energy usage of the biotic community. Therefore, they hypothesize that available energy limits species richness.

The species-diversity concept may be applied to within-ecotope species abundances, multistand abundances, or biome abundances. These have been called alpha, beta, and gamma diversity, respectively.

Implications

Clarence Glacken (1967) remarks that the idea that the world contains the maximum possible number of organisms and that this variety is a good thing is very ancient. The concept of "plenitude" (Glacken attributes the term to Arthur Lovejoy) is found in Plato and other ancient authors, and it continues to form a key element in contemporary conservationists' arguments. Biodiversity is good in itself, but it also has important functions. Most studies of the role of a species in an ecosystem have shown that each has a unique function. It performs some useful task. Species diversity is not some sort of trivial decoration on the cake of nature. Diversity is a measure of the vigor, health, productivity, and beauty of an ecological system. It is an essential attribute of ecosystems.

Although each species has value because it represents a unique track of evolution and selection, species are not static. They change constantly in response to new selective forces in their environment. This process may create problems for human well-being. For example, newspaper reports frequently comment on the emergence of infectious diseases that were formerly thought to be under control. Malaria, dengue, Ebola fever, tuberculosis, AIDS, and other diseases have emerged or reemerged as public health problems. The Centers for Disease Control and Prevention report that deaths from infectious diseases rose 58 percent between 1980 and 1992. Returning to species evolution, it has been reported in these news sources that excessive antibiotic use appears to be the driving force behind the spread of penicillin-resistant streptococcus bacteria. Although the antibiotic use had a short-term advantage, it caused a longer-term problem. How can a balance between excessive control of the pathogen and the costs of some level of ill health be achieved?

The manager of organisms needs to remember that it is always easier to work with organisms, accepting a degree of competition as a cost. It is often more effective to alter the environment where the organism occurs rather than trying to destroy the pest through poison. A quick-fix, short-term solution may merely postpone the problem and make it more difficult to solve in the future. Absolute control almost never is achieved. Rather, our aim is to find an acceptable balance between the costs of the disease and the benefits obtained from control.

In many communities some species are engaged in another kind of balancing act. On one hand, they represent the unique products of selection. On the other hand, they play an essential role in the community by being redundant members of a species complex. Redundancy is an important process. It means that a key function is maintained by the capacity of other elements to continue to perform that function if the principal organism is lost (Moffat 1996). Probably the function will not be performed as well, but the show goes on.

In human-built systems, redundancy is routinely encountered if system performance is essential. An example is a spaceship of which an oxygen-purification system is a fundamental part. Since an astronaut dies if oxygen is unavailable, the system has a back-up. But given the danger of oxygen deprivation, the back-up itself has a back-up and, as a last resort, individual oxygen cylinders are carried on board. Species play similar roles to this sequence of redundant oxygen systems in the spaceship.

The intrinsic and instrumental values of species are so obvious that one would think all humans would recognize them. Indeed, throughout most of history humans have related closely to plants and animals, even claiming joint ancestry and family relationships with them. Humans regularly have given animals human qualities. In modern, urban, mechanized society, however, such relationships are mainly invoked to amuse small children. The adult view is that humans and nature are separate and that humans have dominion over nature. Nature exists only to serve human welfare. In this context, an absurd question, such as how one can equate the existence of one species of small owl to the livelihood of a logger and his family, sounds reasonable. Such questions indicate a lack of understanding of how natural systems are structured and how they function, as well as ignoring time and human adaptability.

As the human population has grown and individuals have demanded a life style similar to that of the richest among them, they have created markets which can be satisfied only by pushing the resource base harder, being less tolerant of sharing production with other creatures, and being more willing to take chances with erosion, land exhaustion, and species loss. The agronomist who claims to "feed the world" might expand his or her thinking to include the world of fellow organisms who share the planet and not focus exclusively on meeting the demands of human beings.

In my opinion, however, the opportunity for adjusting the rate of environmental disturbance, species loss, and resource destruction has passed. We have reached a point where the pressure has become visible in place after place. Calls for increased production become louder and more strident. The exploding human population demands not only more but more per capita. The conservationist responds with dire warnings of a species holocaust where humans destroy everything in a misguided belief that somehow they can overcome "the problem" and survive the crisis intact. The extinction rate of species is probably scores per day. Today's mass extinction of the biota by humans is of the magnitude of the extinction events that occurred in the past.

I believe that the conservationists' apparent stridency is merely commonsense ringing of the alarm bell. But I do not think that we will hear the bell. Our culture closes our ears to it. For this reason it is essential to increase the numbers of conservationists and actively practice environmentally sound ways of living. It is essential that conservationists create a diversity of ways to resist the forces of destruction at work in a thoughtless and uncaring society.

In this context it is useful to repeat the message. It has been suggested that the biota has important instrumental value and that the loss of species may mean loss of future resources or medicines. Although these arguments are probably true, they miss the point that each species is a unique product of evolution and as far as we can tell serves a useful purpose in ecosystems. Species have not evolved without reason. They are the products of selection and represent adaptive solutions to environmental challenge. As such, all species have intrinsic value and deserve our respect, consideration, and protection. We must learn to live within limits of our own making. It is not that we cannot exhaust the Earth; it is that we should not do so. To understand this rule and to apply it personally is a mark of maturity and civilized behavior.

Readings

Grumbine, R. E. *Ghost Bears; Exploring the Biodiversity Crisis.* Covelo, Calif., Island Press.

Kellert, Stephen R. 1996. *The Value of Life: Biological Diversity and Human Existence.* Covelo, Calif., Island Press.

Lawton, J. H., and V. K. Brown. 1992. "Redundancy in Ecosystems," pp. 255–70. In E. D. Schulz and H. A. Mooney, eds., *Biodiversity and Ecosystem Function.* Berlin, Springer.

Moffat, A. S. 1966. "Biodiversity Is a Boon to Ecosystems, Not Species." *Science* 271: 1497.

12

Primary Production and Decomposition

One of the most important properties of the ecosystem is its capacity to produce organic material. All living organisms depend upon this property. It is the source of human food, and the scientific agriculturalist is focused almost entirely on how to manage the production process. Production is the input side of this system process. It is balanced by the process of decomposition or organic matter breakdown. Production and decomposition form one of the sets of dualistic relationships that characterize the natural world.

Production

An ecosystem of simple structure, such as a corn field, has a characteristic production pattern over the growing season (fig. 12.1). At the beginning of the season, the system consists of bare earth. After planting, the corn plants grow rapidly and reach maximum growth in the late summer. The plants then become senile and eventually die. The herbivorous animal populations in these fields follow a similar pattern, feeding on the growing corn plants and then dying, migrating, or going into a resting state as the plants become senile and dry out. During the nongrowing season the field lies dormant, with standing dead plants indicating the past crop.

If we harvest the corn at the peak of its growth, we can measure its production. In this case, we are measuring the production of the entire plant. The farmer is concerned about the production of grain, and he or she measures only the increase in corn grain. Each part of the plant has a unique production curve, as does each animal population feeding on the corn.

The product that is present and measurable at the peak of the production curve represents the product of photosynthesis by the corn plants minus any losses that occurred during the growing season. These losses

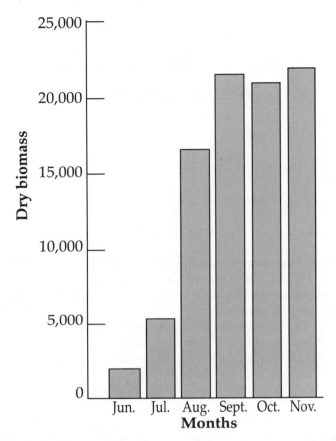

Fig. 12.1 Dry-matter biomass of maize (kilograms per hectare) in irrigated fields in northern Spain. Data from an unpublished master of science thesis by Fermin Cerezo, Institute of Mediterranean Agronomy of Zaragoza.

might involve tissue eaten by herbivores, leaves which were low on the stalk that died and fell to the ground, and the energy (following the second law of thermodynamics) that the plant required for its own maintenance. This latter quantity we usually call the respiration cost (R). The product measured at the peak production period is called net primary production, or NPP. It is "net" because it is what is left after the losses have been subtracted from the total production. It is "primary" because it represents the first step in the food chains and webs that support all the other members in the ecosystem.

The total primary production is called gross primary production, or GPP, and it represents the conversion of sun energy into organic material by the plants through the process of photosynthesis. Thus, the formula for production is: GPP − R = NPP.

Limiting Factors to Production

The photosynthetic process requires a variety of resources other than light energy. They include water, carbon dioxide, various nutrients, such as nitrogen, and proper temperature for growth. As a consequence, productivity varies with environmental conditions and resource availability. We will consider the role of two of these environmental factors.

Light. An example of the interaction between the vegetation and the light environment is presented in figure 12.2. This graph contrasts three species of plants that typify three climates. Each species requires light energy as a basic resource. At low light intensity or at night there is not sufficient light energy to allow photosynthesis to occur. Nevertheless, at these times the energy cost in metabolism continues. As a consequence, during these periods the plant uses energy stored in organic compounds that were produced during the daylight hours. We say that they lose energy in respiration because the exchange is often measured as the release of carbon dioxide to the atmosphere. The maintenance costs of plant life go on both night and day and make up the respiration cost. As light intensity increases photosynthesis becomes possible and production exceeds maintenance costs. In the species from the rain-forest floor the production rate is never very high when compared to that of the other plants. In this species, adapted to life in the dense shade of a tropical forest, the rate of photosynthesis is constant at low light intensities, indicating that the plant can't respond to more light energy. In contrast, the temperate weed species, whose curve is shaped similar to that of the rain-forest species, can utilize a greater range of light intensities and has a higher production rate. Finally, the desert species is adapted to high light intensity and, if sufficient water is available, can increase its production over the range of light intensities shown on the graph. It does not reach a point of light saturation where production levels off, but its capacity to produce depends on another environmental factor: water.

Water. Primary production varies with the moisture level (fig. 12.3). The figure provides data comparing two widely different environments: the Great Smoky Mountains of Tennessee and North Carolina and the Santa Catalina Mountains of Arizona. In the Smokies the tree life form dominates the production spectrum at high moisture levels. Under dry conditions herbs are more important. Shrubs are unimportant under high moisture conditions but increase in importance as conditions become drier. In Arizona the pattern is different. Trees are dominant under moist conditions, but as conditions become drier shrubs dominate herbs.

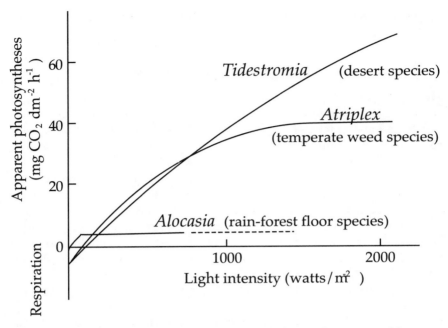

Fig. 12.2 Comparison of the relation between photosynthesis and respiration of three species representative of three different types of ecosystems. Data from Salisbury and Ross 1978.

These patterns are partly explained by the plants' response to stress. Plants cope with stressful conditions by storing energy and nutrients. When production resources are likely to be inadequate, it is an advantage if there is stored organic material, which can be mobilized quickly in response to improved growing conditions. In deserts, shrubs can resist drought and respond to irregular rain conditions better than herbs can because of their larger and deeper root systems. But when growing conditions are less stressful and moisture levels are adequate, the species of plants that can compete for light energy as a resource have an advantage. Therefore, under such conditions trees dominate the production spectrum because they grow taller than shrubs.

A consequence of the interaction of plants and their environment is a pattern of widely different production levels in different ecosystems. On a regional scale (fig. 12.4), the highest levels of organic production in North America are in the southern United States, where the growth period is long and moisture conditions are relatively favorable. Note how this map correlates with the ecoregion map of Bailey (see figure 6.3).

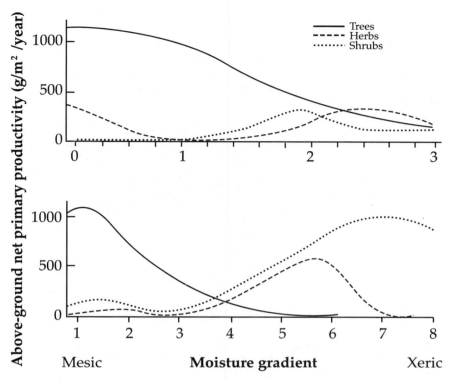

Fig. 12.3 Trends in annual above-ground primary production along moisture gradients in the Great Smoky Mountains, Tennessee and North Carolina (top graph from Whittaker 1966) and the Santa Catalina Mountains, Arizona (lower graph, from Whittaker and Niering 1975). Tree, shrub, and herb vegetations are plotted separately against a gradient ranging from mesic to xeric conditions. The productivity data are in grams per square meter per year.

Decomposition

The energy in the net primary production is available to the other members of the biota through the food web. The food web has two major flow paths. One goes to organisms that consume green, living plant tissue. These organisms are called herbivores or biophages and include cattle, rabbits, birds, and many insects. In a typical terrestrial ecosystem, about 18 percent of the production flows are in this part of the food web. In contrast, in aquatic systems with large plants, such as kelp, the percentage of herbivory is about 30 percent and in systems with floating algae it is much higher, 51 percent (Cyr and Pace 1993). The rate of herbivory and the herbivore bio-

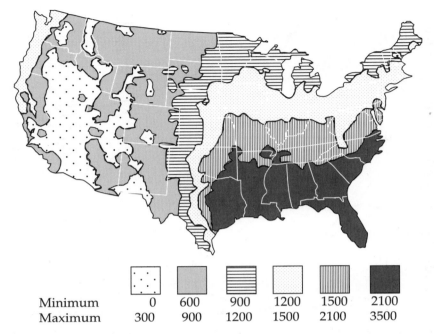

| Minimum | 0 | 600 | 900 | 1200 | 1500 | 2100 |
| Maximum | 300 | 900 | 1200 | 1500 | 2100 | 3500 |

Fig. 12.4 Average annual net primary productivity in the United States. Production data are in grams of dry matter per square meter per year. Data from Lieth and Whittaker 1975.

mass are highly correlated with primary productivity in both terrestrial and aquatic systems.

The second pathway leads to organisms that consume dead and decomposing organic material. These are called saprophages or decomposers. About 80 percent of the terrestrial primary production flows through this part of the food web. The process of breaking down the organic material requires physical disassembly (which in the case of a large log is a long-term and difficult task), followed by chemical disassembly into chemical molecules. The molecules can be reabsorbed by the roots and reused by the plants. Often insects begin the process of physical decomposition, and bacteria and fungi complete the process of chemical reduction.

Production and decomposition must be balanced. If decomposition exceeds production, the system runs down and eventually dies. If production exceeds decomposition, the system stores the energy and materials and grows in size and mass. Ultimately, the system structure is altered, the production process is changed, and growth and decomposition find a new balance. An example of the latter effect is the accumulation of partly decomposed organic material, called peat, in swampy habitats. As mosses and grasses grow in the swamp or mire system, the dead plant material cannot decay rapidly enough to balance the rate of production because water prevents oxygen from reaching the dead organisms during the growing season and cold temperatures reduce the rate of decay during the winter. The consequence is an accumulation of dead organic matter. This material becomes compressed over time and forms a coallike material that can be burned. The traditional Irish or Scottish turf fires that flavor Irish and Scotch whiskey come from peat fuel. If the peat is not removed by humans, an equilibrium is reached between the rates of production and decay conforming to the water levels and nutrient conditions of the site.

The process of organic matter break-down in the soil, water, or sediments is a control point in ecosystem dynamics. Ecologists speak about a balance in community production and decomposition, or the P/R ratio, as an indicator of the state of the system. Although balance is never achieved in a static sense, there may be a dynamic balance, with accumulation of product one season or year and a decline in the system stock another year. Systems with a P/R ratio below or above 1 tend to return to a balance over time, unless they are continuously disturbed.

Implications

In examining production and decomposition we have encountered once again a balanced dualism, which is a fundamental aspect of our perspective of nature. But Euro-American culture, which focuses on production, seldom considers it as one side of a relationship. Rather, production is usually treated as a distinct and separate process which, given adequate resources and human will, can increase indefinitely. A large share of scientific and technical research is devoted to assuring that this assumption is realized.

From our discussion, however, we know that continuous increase in production is unlikely, even in economic terms. Continued production results in accumulation of product that creates new structures, which in turn change the relation of production and decomposition. Thus, unlimited production necessarily feeds back to change the system structure and function, eventually causing a decline in the production rate. Here again, Aldo Leopold's dictum that an action is right when it maintains the stability and integrity of the biotic community holds true. In this instance we can interpret stability of the biotic community as meaning stability in the rate of production balanced by decomposition, and integrity as meaning maintenance of the structure of the production-decomposition system.

Productivity varies with the maturity of the ecosystem. Mature systems use most of their energy fixed through photosynthesis in the maintenance of the system. As the system grows older there is a build-up of organic biomass that requires energy to be maintained. Eventually the maintenance cost becomes so great that the system grows senile and is only renewed by a catastrophic fire or attack by insects. In this situation the forest oscillates between mature and immature states over the landscape and over time.

In the mature system little production is channeled to herbivores. Most goes to the decomposer parts of the system. As a consequence, old-growth forests are not very attractive to a production forester, but they are attractive to the conservationist who seeks balanced systems over long time periods.

Humans recognized these differences early in human history and settled in habitats where disturbance was frequent and the ecosystem remained in the rapid growth phase. A river delta that was periodically flooded or a savanna which regularly burns were potential sites for early human habitation. The invention of agriculture was essentially the manipulation of the production process through human control of the intensity and timing of these disturbances. The environment was manipulated through burning, scratching the soil surface with digging sticks, or channeling flood water. These are all simple steps toward management of production and assuring a predictable food supply, which in turn allows settled life and the construction of buildings. Modern agriculture manages the environment through plowing, control of water, addition of essential nutrients through fertilizer, and management of competition with other herbivores through pesticides.

The success of agriculture has led humans to disturb much of the Earth surface for the production of food and fiber crops. Production rates have expanded to dizzying heights by applying fossil-fuel energy to the process. The consequence of this evolution has been hubris, an attitude that humans are outside of the natural process and that production can go on increasing endlessly. Further, by living exclusively in landscapes dominated by agriculture, humans have lost the sense of balance that comes from interaction with mature natural systems. As a result, growth processes have been given high value and maintenance processes low value. Yet production and maintenance are equally necessary to any sustainable system. Both are of equal value. Both deserve respect and regard.

Readings

Goulden, Michael, J. W. Munger, S. M. Fan, B. C. Daube, and S. C. Wofsy. 1996. "Exchange of Carbon Dioxide by a Deciduous Forest: Response to Interannual Climate Variability." *Science* 271: 1,576–78.

Johnson, Edward A., and Kiyoko Miyanishi. 1991. "Fire and Population Dynamics of

Lodgepole Pine and Englemann Spruce Forests in the Southern Canadian Rockies," pp. 77–91. In N. Nakagoshi and F. B. Golley, eds., *Coniferous Forest Ecology, from an International Perspective.* The Hague, SPB Academic Publishing.

Waring, Richard H. 1989. "Ecosystems: Fluxes of Matter and Energy," pp. 17–41. In J. M. Cherrett, ed., *Ecological Concepts.* Oxford, Blackwell.

13

Ecological Succession

Ecological systems are disturbed by many natural environmental events, such as tornadoes, hurricanes, fires from lightning strikes, and diseases. When these disturbances exceed the capacity of the species to adapt and recover after the disturbance has ceased, the biota is partly or wholly destroyed and the physical environment may be changed through substrate erosion and nutrient loss. Once the disturbance stops, however, plants and animals reinvade the site and an ecological system is gradually reestablished. The process of reestablishment of the ecosystem is called ecological succession.

Historical Development of Succession Theory

Ecological succession has been a dominant theme in American ecology, if for no other reason than that Americans are faced with a highly disturbed landscape that is constantly being rearranged through human manipulation. The theory of ecological succession was also one of the first grand theories of the discipline and has been the subject of an enormous amount of debate and study.

An early and significant description of ecological succession was made by Henry J. Cowles, professor of botany at the University of Chicago. Cowles observed that the shore of Lake Michigan was organized into a series of sand dunes of successively older ages (Cowles 1901). As one walked away from the lake, the plants on these dunes formed a sequence that indicated the history of the vegetation. Eventually one reached forests consisting of basswood and several species of oaks, which were characteristic of the regional vegetation. The sequence on these dunes formed a series—a succession of stages (fig. 13.1). These stages represented a series of eco-

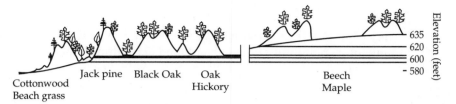

Fig. 13.1 A profile of the sand dunes at Gary, Indiana, and the Indiana Dunes State Park, showing the increased stability of the surface over time. Based on Olson 1958.

logical communities which appeared to replace one another. This was the key idea in the theory of succession.

Another American botanist, Frederic Clements, proposed that ecological succession was more than a process of vegetating the land. Clements focused on the end point of the process. He proposed that succession was a development that led to a mature, stable vegetation characteristic of the regional climate. He called this mature stage the climatic climax. Clements also proposed that each sequence on a site represented a stage in a universal process of development and that each moved toward the same end point appropriate for the region or biome, albeit at different speeds. In Clements's view, there was a successional force that not only led to revegetation of disturbed lands and water but also moved to a single climax for each of the climatically shaped biomes discussed in Chapter 7.

Clements's (1916) concept of succession was so inclusive, so deterministic, and so orderly that it had a major influence on American ecology. It provided ecologists with a way to organize into meaningful patterns the disorderly assemblages they observed on land abandoned to farming, recently cut over or burned lands, and the heaps left from mining and industry. Although individual ecologists, such as Henry Gleason (1927), disagreed with specific parts of Clements's interpretation of vegetation processes, general doubts about his theory only arose after the great drought in the 1930s (Tobey 1981). Contrary to Clements's theory, the western grassland vegetation did not recover from drought through a sequence of stages leading to recovery of a stable state. Rather, the patterns of recovery were contingent on the environmental conditions of the site and the species that invaded it. Chance played an important role in the process of recovery, and in many instances the supposed climax was never reestablished.

After the Second World War Clements's theory of succession ceased to dominate American ecology. Ecological succession is no longer viewed as a deterministic process. Now the term refers to the recovery of vegetation and ecosystems from disturbance and does not involve the progressive development of vegetation through a sequence of stages toward a fixed end point.

A Modern Interpretation

The recovery of an ecosystem destroyed by glaciation, fire, or the develop-ment of the biotic community on an abandoned agricultural field repre-sents the ecological process of succession. S. T. A. Pickett and his col-leagues (1987) describe the causes of succession: (1) an open site must be available; (2) species must be differentially available at this site; and (3) these species must have different abilities for dealing with the site envi-ronment and each other. These conditions provide the foundation for recovery.

The process involves several steps. First, there must be dispersal of the biota from the landscapes surrounding the disturbed site or growth from seeds and larvae that remain in the soil. The second step requires estab-lishment of these invaders. Third, the biota that become established must obtain sufficient resources to grow, mature, and reproduce so that viable populations develop. After completion of these steps, one can conclude that the site has been occupied.

As a consequence of this biological recovery process, the initial in-vaders of the site are usually species that disperse readily, germinate under difficult conditions, grow rapidly, and reproduce easily. Frequently they are weedy species that rely on wind, water, or other organisms for their disper-sal and are widely distributed over the landscape. Such pioneer species dominate the biotic community of a disturbed site during the first years of recovery. Chance plays an important role in determining which species ar-rive and become established.

Not all the species that invade the site early in succession are rapid-growing, weedy species. Among the arriving species are some that grow slowly, establishing deep root systems that store nutrients and energy in below-ground organs. These slower-growing species include members of the grasses, conifers, and hardwoods. Eventually, these latter species in-crease in size, garner resources, outcompete the first pioneers, and begin to dominate the stand. It may appear that a new community has formed and replaced the pioneer community, because the site has become shrubby or covered with small trees. Actually, the pioneer species often are still pres-ent, they are only less conspicuous.

The biota also acts upon the physical-chemical environment by adding organic matter, by digging tunnels and holes, and by accumulating nutri-ents. The changed environment is hospitable to other species that can now invade the site, establish themselves, and grow. The productivity, mainte-nance cost, and biomass of the community increase (fig. 13.2).

To the uneducated eye, the stand appears to go through a sequence of stages in which the life form of the vegetation changes. For example, in the

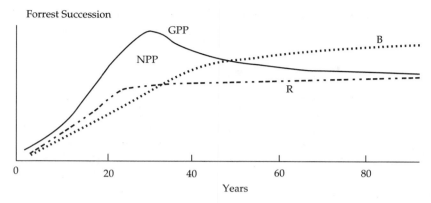

Forrest Succession

Fig. 13.2 The pattern of ecological change in the temperate-forest biome after land has been allowed to undergo revegetation. The patterns show the rise in gross primary production (*GPP*), which reaches a maximum early in the succession, biomass (*B*), and maintenance costs (*R*). The difference between *GPP* and *R* is the net primary productivity (*NPP*).

southeastern United States, where Henry J. Oosting made his classical description of succession of abandoned agricultural fields, recovery involves, first, pioneer herbs, which are mainly composites and legumes. This stage is followed by perennial grasses, dominated by broomsedge (*Andropogon spp.*). Then the pines become dominant, and finally a hardwood forest of oaks and hickories is established. Actually, many of the species invade the site at the same time, but it takes ten to twenty years for some of the trees and grasses to become visible to an observer, and two hundred or more years before a mature oak-hickory forest is present.

Eventually, a community is reestablished that contains many of the species that can live under the soil, climatic, and physical-chemical conditions that characterize the biome or landscape. It is unlikely that the new community will contain exactly the same species which were present before disturbance. Chance plays a role in what species invade and become established. The regional climate, the local geomorphic and soil conditions, and the size and condition of the disturbed area determine what broad type of community will reoccupy the site. Usually a hardwood forest will be reestablished in a hardwood forest landscape. This is not always true, because some communities are relict communities which developed under environmental conditions that no longer exist. When a relict is destroyed, it will not recover. Moreover, a disturbance may be so severe that the soil is removed or the topography changed. In this instance, succession is called primary succession because the biotic invasion must begin on a physical surface that has never been modified by living organisms. Often primary succession, seen for example on rock outcrops in glaciated landscapes, is carried out by lichens and mosses. Slowly, these plants accumulate a soil that will support grasses, herbs, shrubs, and even trees.

Ecosystem Stability

Any view of a modern landscape from the perspective of ecological succession would lead us to conclude that most of the natural, unmanaged ecotopes in these landscapes are in a state of relatively rapid change. There are almost no mature ecotopes left in the modern landscape. Mature systems are usually only found in parks or reserves, where they can be protected. To the consternation of the managers of parks and the general public, however, these protected ecotopes are also changing. They are burned, overgrazed by protected animals, stricken by drought and flood, blown over by tornadoes, invaded by starving peasants and their livestock, insulted by air pollution, and so on.

Are these mature systems end points of development? Should they exhibit stability of structure and function? Or are they merely chance collections of long-lived species, which we have arbitrarily labeled "mature," "climax," or "stable"?

The question of whether ecosystems maintain a stable state is of fundamental importance in ecology. What exactly do we mean by stability? The dictionary definition emphasizes enduring, fixed, steady, constant properties. The ecologist finds the question difficult, because there is little agreement on the significant properties of an ecosystem that must remain fixed, steady, and constant. Properties of the physical environment vary enormously depending on the sphere we are speaking about. The biota is dynamic and highly responsive to its environment and is also variable over space and time. Which properties can we use to address the issue of ecosystem stability?

The evaluation of the stability hypothesis also is difficult because there are few undisturbed ecosystems in which the theory can be tested. Most of the world's systems have been disturbed through air pollution, human management, species extinction, and so on. Where one can observe natural landscapes or waterscapes, as in the remaining Amazon forest, the forest life form appears to be stable. Even small-scale shifting agricultural plots are rapidly reforested. Yet if one's definition of stability requires recovery of the same species in their former ratios of abundance in the new community, it is unlikely that even the Amazonian forest would be defined as stable. Like Heraclitus's river, which flows on and is never the same, no new ecosystem will repeat exactly the biotic composition or physical-chemical environment of a former system.

Ecologists have used experimental systems to study stability. David

Schindler and his associates at the Experimental Lakes project in Canada divided small lakes into two parts, disturbed one part experimentally, and then watched the patterns of response. Among these experiments was the purposeful acidification of a lake (Schindler et al. 1985). The investigators found that this treatment resulted in a change in the species dominance of the lake, but the lake function was robust and resisted change. If we focus on the lake biota, we conclude that the ecosystem was not stable. If we focus on processes such as productivity, then we conclude that the lake was stable and that it was the variability of the species diversity that maintained function. In systems language we would say that the redundancy of species allowed one species to replace another that could not cope with the acidic conditions.

Thus, the answer to the question "Are ecosystems stable?" is "It depends." It depends on the type of system, the time frame, and our criteria of stability. Systems are fundamentally dynamic, because the physical environment and biota are dynamic. Yet the time required for significant change in the physical environment is usually greater than that for biotic change. Survival of the biotic community depends on its dynamism—the creative capacity of the biota to adapt—and change allows the ecosystem to respond to external pressure and maintain life form and function.

Implications

Consideration of succession and stability leads us toward recognizing a yogalike or "akidolike" response in ecosystems (Thorne and Pitz 1988). The combined species that make up the biota of the system are in continual interaction and adjustment to each other and the environment. This is what we mean by a dynamic response system. There is no purpose, no direction, no static end point that the system is moving toward; rather, there is a process of living, adapting, interacting, and surviving. If a system is overcome by disturbance, then new species invade the disturbed site and another system is created. This new system will be different in detail from the former one, but it too will fit the climate and the soils of the site and will be made up of the biotic species that exist in the region. The human interpreter can focus on the chance occurrence of a species or on the repeated pattern of the life form. Both are present.

Conservationists have frequently tried to maintain ecosystems in a desired state. For example, America's national parks were managed to retain the condition observed when they were first seen by Europeans. Park managers found this a difficult objective that grew more problematic with time. The natural processes of growth and development lead to the accumulation of fire-susceptible vegetation, and therefore to the chance of catastrophic fires. For example, the 1988 fire in Yellowstone National Park burned 292,000 hectares (Turner and Romme 1994). Landscape architects have also tried to design with nature and create ecologically stable landscapes. Again, they found that successional dynamics changed the form of the community they created. These landscapes required costly management to maintain the design. In both examples, nature is treated as a mechanical system with fixed operations. The system is expected to perform in a consistent fashion. As we know, nature is not a machine. Rather, it is ever-changing, so that the conditions of the moment are not likely to be exactly repeated in the future.

What is the objective of the conservationist? Is it to manage for stability or dynamism? Is it best to step back and let nature take its course or to intervene continually to achieve a desired goal? Do we know enough about the consequences of managing wild land to predict our chance of successful intervention? The environmentalist would likely allow nature to act dynamically, reducing human interference to the minimum. The reasons for this decision are twofold. First, the biota has evolved to fit the conditions of the site, and evolution and adaptation will continue if the biota is left to develop without interference. We may count on nature operating according to its time-honored rules, but we can't predict the outcome. Second, it is doubtful that we know enough to design and operate a stable system. Around us are abundant examples of our inability to maintain the stability of the ecosystems created for our own use—agricultural fields, forest plantations, fishing areas, and so on. We know even less about natural systems. Possibly, we will reach a point when our knowledge is sufficient, but there is little evidence that it is sufficient today.

Readings

Golley, Frank B. 1977. *Ecological Succession*. Benchmark Papers in Ecology, vol. 5. Stroudsburg, Dowden, Hutchinson, and Ross.

McMahon, J. A. 1981. "Successional Processes: Comparison among Biomes with Special Reference to Probable Roles of and Influences on Animals," pp. 277–304. In D. C. West, H. H. Shuggart, and D. B. Botkin, eds., *Forest Succession: Concepts and Applications.* New York, Springer.

Pickett, S. T. A., S. L. Collins, and J. J. Armesto. 1987. "Models, Mechanisms and Pathways of Succession." *Botanical Reviews* 53: 335–71.

Cluster 3

The Population and the Individual

At this point we move from a focus on land and water systems to a different but equally important set of systems. The set forms a new ecological hierarchy that includes species, metapopulations composing the species, local populations, and individual organisms. These biological units interact with their environments to form ecological systems in the same way as watersheds and ecotopes. But our focus shifts from the energetics and biogeochemistry in spatially defined systems to the biological properties and interactions of the organisms in the systems. Because these organisms are frequently our size or smaller and share many traits with us, and because we can observe, hunt, or domesticate them, we approach them with a sense of familiarity. Although we could treat them in the same way as we treated the flows and storages in ecosystems, we do not. Instead, the environment is expressed as a resource gradient or a limiting factor to the organisms. Our interest is in knowing how the organism copes with and utilizes a dynamic environment.

The population-individual hierarchy can be examined from a structural, functional, developmental, and relational viewpoint, as were the land and water systems. But the ecologists studying these systems seldom use this language. Rather, they speak the language of biology, economics, sociology, and psychology-behavior. But there are more subtle differences between these approaches to ecology. It is unusual for a person to go beyond the view of land and water as a resource or as an attractive landscape to a deep emotional involvement with them. We treasure the writing of authors such as Sigurd Olson and John Muir because of their capacity to express keen perception and love of nature. In contrast, it is common for people to form a profound attachment to organisms. We love pets and livestock, we thrill to see wild animals, and some of us live for the hunting season when we can return to the forest or prairie. The field ecologist who studies plant or animal populations almost always has this intensely emotional relationship with the creatures he or she studies. The individuals are given names and become members of the family. One of my favorite examples of this type of field ecologist is Frank Frazier Darling, who made field studies of the natural history and behavior of red deer in Scotland, but there are many others to choose from. This attitude brings the set of concepts in Cluster 3 closer to us; we can identify with populations and individuals. Then too, the lessons have direct application to humans.

Cluster 3 consists of five chapters, beginning with the population and then moving to the individual. The individual is the finest level of detail that we will examine in this book. If we decomposed the individuals into the next level of subsystems in genetics, physiology and behavior, we would be entering the field of conventional biology. Ecologists use biological information and concepts to explain the ecology of individual organisms, but seldom do they carry out biological, in contrast to ecological, studies. Of course, some biologists are also ecologists and readily move across the ecology-biology boundary. Fields of study have fuzzy boundaries, just like ecosystems.

14

The Population as a Demographic Unit

The word *population* has both a popular and a technical meaning. Ordinarily we use population to mean an aggregation or grouping of people or organisms. What is left open by this definition are the criteria used for selecting what is or is not in the group. In an ecological context, population is usually applied to an aggregation of living organisms which are members of the same species, share a genetic constitution, exchange genetic materials, and live in the same habitat.

These appear to be straightforward, unambiguous criteria, but that is often not the case. For example, Paul Ehrlich and his colleagues and students (Ehrlich and Murphy 1987) studied the checkerspot butterflies on a nine-hectare grassland on a serpentine ridge in northern California for twenty-seven years. The serpentine soil has unusually high quantities of magnesium that restrict the plants growing on the ridge to grasses and herbs. The habitat is continuous and there are no barriers to butterfly flight, yet the nine-hectare grassland has been occupied by three separate demographic units that qualify as populations by the criteria presented above. Fewer than 5 percent of the individuals marked in one unit were recaptured in another. These three populations of butterflies have their own separate dynamics. For example, one of the populations went extinct in 1964, was reestablished—presumably from one of the other populations—in 1966, and then went extinct again in 1974. In 1996 all butterflies were absent from the grassland, but in 1997 a small population was reported.

The casual observer walking across the nine-hectare serpentine grassland might observe the butterflies and conclude that there was a population of checkerspot in this small habitat. Without Ehrlich's intense and sustained interest in these organisms over the years, we would never know that they formed distinct populations, with a low rate of transfer of individuals between them but apparently with the capacity to reestablish a population that goes extinct. Thus, the three populations were individual and distinct ecological entities, yet they made up a higher-level entity too.

This higher entity is the combination of the three populations occupying the serpentine grassland. It may be separated from the next collection of checkerspot populations by many miles of unsuitable habitat. We call this collection of populations within a discrete area a metapopulation. In this case, the prefix "meta" means going beyond a single population to embrace a group of populations. In turn, the collection of metapopulations across the biome or continent makes up a species.

The size of the habitat available to the butterflies is not the only factor required to assure that this metapopulation is viable. A large homogeneous habitat may not be able continuously to support the butterflies, because a disturbance such as a fire across the nine hectares could negatively affect all the populations at the same time. Topographic diversity is required in addition to size of area. For example, successful hatch of butterfly eggs depends on local environmental conditions. North-facing slopes usually are preferred by the butterflies, but in wet years these north slopes are not as attractive, and the south-facing slopes then contribute to egg survival. In the event of a fire, topographic diversity might provide a refuge where a population could survive. One hopes that a refuge population might reestablish this species on the serpentine ridge.

Distribution

These observations show that an important feature of populations is their distribution. Population distribution can be viewed at several spatial scales. At the biome scale, the distribution of a wide-ranging species is recorded on range maps which show where the species might be expected to occur in suitable habitats throughout the biome (fig. 14.1). It is not likely that all the habitats are occupied and there may be many more hectares of unsuitable habitat than there are suitable habitat. Members of the species could interbreed across the biome, at least theoretically, even though the individuals are spread over a large geographic area. This is the distribution presented in the maps in bird or wildflower identification manuals.

At a finer landscape scale, a metapopulation occupies a landscape or an ecotope. Transfer of genetic information through migration of individuals can occur between subunits of the metapopulation, but the subunits are not contiguous in space or time. At the finest scale are local populations associated with specific ecotopes.

This scheme does not fit all populations. For example, the three populations of checkerspot butterflies mentioned above occupied a single nine-

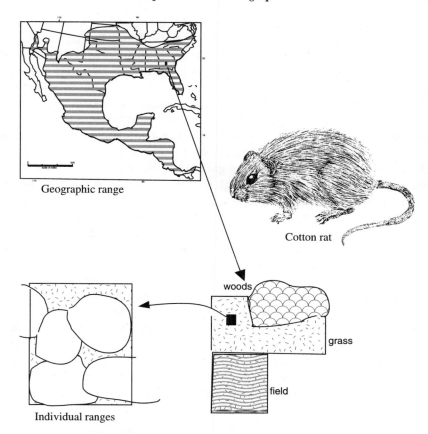

Geographic range

Cotton rat

woods

grass

field

Individual ranges

Fig. 14.1   Distribution of the cotton rat (*Sigmodon hispidus*) at three landscape scales. The upper map describes the range of the species, the lower-left map the range of individual animals, and the lower-right map the habitat of a single population.

hectare ecotope of serpentine grassland on a ridge. The butterflies evidently view this ecotope as more varied than it appears to the ecologist. From a butterfly's perspective, there seem to be three quite distinct ecotopes on the ridge. It often takes a lot of time and effort to learn to see the environment as it is seen by another species.

Density

Besides distribution, the second piece of information required for an ecological study of a population is its density. Density means the number of organisms per area of habitat. Actually, many parts of a region or landscape may not provide the conditions required by the organism. Therefore, we often calculate an ecological density to make the density estimate more representative of the resources available for the species. Ecological density means the number of organisms per unit area of habitat where the species potentially could live. For example, if a population is restricted to a specific kind of plant growing in patches within a grassland, it is not very useful to report the density as the number per total grassland area. Density related to the area of habitat occupied by the host plant would be more relevant. Further, if an animal population is studied in grassland patches in a forest matrix but sometimes occurs in the forest, it is necessary to specify that the density estimate refers only to the grassland patches, not to the combined forest and grassland areas. Although abundance is almost always reported for a population, often it is not clear how the density relates to the habitat.

The study of population density is a special field called demography. We can examine the population from a variety of points of view other than its demographics, but from an ecological perspective this approach focuses on a basic feature of the natural pattern. Ecologists usually begin by counting numbers of organisms and then use these data to build hypotheses about the relations between the organisms and their environment.

Population density is often highly variable over space and time. Habitats of similar size and character frequently contain widely different abundances of an organisms. A graph of population density over time is almost always represented by a fluctuating line, with a few periods of unusually high or low density and many intermediate levels between these extremes (fig. 14.2). A regular or a cyclic pattern of density is unusual and attracts an ecologist's attention immediately.

Population Growth and Regulation

The variable line representing population density over time implies that the population has the capacity to increase when conditions are favorable and then decline during unfavorable periods. Density is the result of the interaction between forces for increase and forces for decrease. The equation for density contrasts these two opposing forces:

$$\text{density} = (\text{births} + \text{immigration}) - (\text{deaths} + \text{emigration})$$

A population that grows has an excess of the force for increase over that for decrease. This force may be due to the births of new individuals or to movement of organisms into the population from other populations through immigration.

If we chart the growth of a population, we often will find a typical pattern. The graph of growth will have an S shape, with an initial period of slow growth, a middle period of rapid growth, and a final period of slow

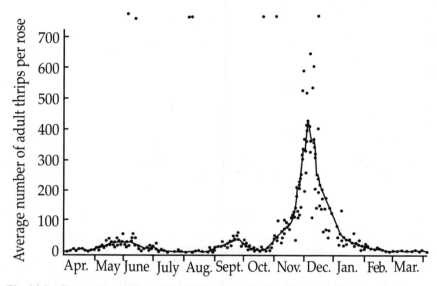

Fig. 14.2 The number of rose thrips (*Thrips imaginis*) throughout a year in Australia. From Andrewartha and Birch 1954.

growth (fig. 14.3). In the middle period the population is increasing at its maximum rate under those particular conditions of the environment and genetic structure of the population. Growth during this period is called the intrinsic rate of natural increase and is given the symbol r. As the population grows at the intrinsic rate of natural increase, it eventually meets limitations of the environment and growth begins to slow down. Ultimately, growth stops. This point where the environment limits growth is called the carrying capacity of the environment, or K. Population growth may continue and exceed carrying capacity, ultimately destroying the environment, it may decline, or it may stay constant. Constancy is not a straight line but rather the highly variable relationship between density and time that we observe for so many species. The variation is between the limits that we have defined as the carrying capacity for that organism in that habitat.

If the population exceeds the carrying capacity, it usually consumes and ultimately destroys the resources on which it depends and then crashes. In this case, a J-shaped curve of population growth, rather than the S-shaped curve, is observed. Populations exhibiting the J-shaped curve increase ex-

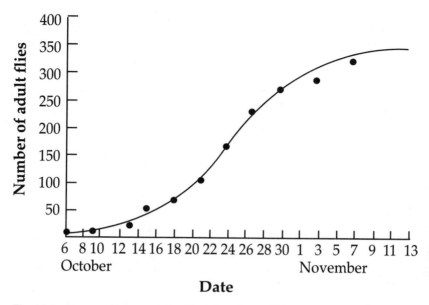

Fig. 14.3 The growth in numbers of fruit flies (*Drosophila*) in an experimental chamber, showing a sigmoid growth curve. Data from Pearl 1927.

ponentially at the intrinsic rate of natural increase until they reach a point where the population exceeds carrying capacity and declines rapidly. At the point of inflection of the curve the population crashes, goes extinct, or persists at a low population level until conditions are suitable for another growth spurt.

In contrast to linear growth, where increase is by the same amount each period (that is, 1, 2,3, 4 . . .), exponential growth means that numbers increase according to an exponent of growth. Commonly we use an exponent of 2 to express this phenomenon, in which case growth is 2, 4, 8, 16, 32, and so on. Exponential growth is explosive because when large numbers of individuals are present the rate of increase is proportionately very high.

The S-shaped growth curve is commonly seen in biology. It fits the growth of the individual body, growth of populations, and other phenomena. It is called the logistic growth curve and was discovered by a Belgian mathematician, Pierre François Verhulst, in 1845. The expression of the logistic growth curve in differential form is expressed as the function of the number of individuals (*N*) at time *t*:

$$dN/dt = rN \left((K - N)/K \right)$$

where dN/dt is the rate of increase of the population and the constants r and K are as defined above.

The tension between r and K defines the relationship of the forces for increase against the capacity of the environment to support those forces. Some species emphasize the growth element in their life history and have a capacity to invade a habitat easily and grow rapidly. These r-strategy species are typically weeds or the pioneer invaders on land that has been abandoned and is being reoccupied by natural organisms. In contrast, other organisms have evolved to use the resources available at carrying capacity efficiently. These K-strategy species grow more slowly, live a long time, and often store resources. In ecological succession we saw species demonstrating both these strategies. The invading species are usually r-strategists, while the species in the mature phase of ecosystem development are usually K-strategists.

Population numbers and distribution are controlled by limiting factors of the environment, as well as by the species' capacity to migrate and adapt. If the population is usually controlled by the climate or another external environmental factor, we say that the population is under independent control, because the controlling factors are independent of the size of the population. A seriously cold winter affects a large or small population in similar ways. An example of this form of control was reported by Andrewartha and Birch (1954) for the rose thrip, a small insect that occurs on roses (see fig. 14.3). In Australia, they showed that the insect population was sensitive to the environmental factors of temperature and rainfall, reaching peak numbers during the Australian summer in November and December. For each 5-degree centigrade temperature rise, the thrip population increased 25 percent. The effect of temperature on thrip increase was three times the effect of rainfall.

In contrast, other populations have the capacity to regulate their populations through behavioral, physiological, and ecological adaptations. In this case, the control is dependent on the size of the population and is called density-dependent regulation. For example, J. J. Christian and D. E. Davis (1964) suggested that in small mammals increasing population numbers generate stress by increasing negative interaction and competition for resources. The stress affects the physiology of the animals, especially the endocrine system. The animal reacts physiologically and adapts to the stress. However, continued excessive stress, which could occur at high population levels, can exhaust the organism's capacity to adapt and results in a decline in its reproductive ability and even in death of individuals. Thus, the organism has a physiological mechanism that causes individual death but helps regulate population numbers.

The carrying capacity concept is very significant in population studies, but it can be misleading if it does not include a time dimension. This is because a population can exceed carrying capacity for a time as it consumes its resource base. In economic terms, this would be analogous to spending one's capital for daily expenses. But in both cases the resources eventually will be used up and the population must decline. A population cannot exceed its carrying capacity indefinitely.

Animal and plant populations seldom become stable or fixed at their carrying capacity. Rather, they fluctuate below this level as factors for increase or decrease vary and as environmental conditions fluctuate. However, the long-term studies that established the concept of carrying capacity, by Paul Errington (1934) in Wisconsin and Iowa, showed that over a landscape population densities of birds and mammals were relatively predictable unless environmental conditions were unusual.

Implications

The concepts of population growth and control are as applicable to the human species as to any other form of life. The population curve of the human species, as a whole, shows a very long period of slow growth and then a period of very rapid growth in the past several hundred years (fig. 14.4). The pattern is thousands of years long, and it is difficult for the average person to extrapolate from the diagram to their personal condition. The problem is made more difficult because the patterns of growth vary greatly from one subpopulation of humans to another. Some of these groups are concerned about survival, others about overpopulation. Finally, human demography is based on changes in the reproduction and death of individuals, both phenomena that are highly personal and strongly regulated socially by humans. Thus, one's point of view about human population growth depends on perspective, gender, and education, and also on one's culture and geographical location. The environmentally knowledgeable person asks the obvious question: What is the relationship between increasing human population and the concept of carrying capacity? Have humans exceeded or will they exceed their carrying capacity? Does this set of concepts apply to the human organism?

To address the last question first, from an ecological perspective there is no reason for human demography to be fundamentally different from that of any other living organism. The principles should be applicable to all populations. However, each species, population, or metapopulation has a unique history that adjusts the general pattern to it alone, leading to the variety of population dynamics we have observed. It is obvious that humans have many special attributes, including the ability to create built environments, and these influence the relationship between carrying capacity of the environment and human numbers. The task, then, is to examine human

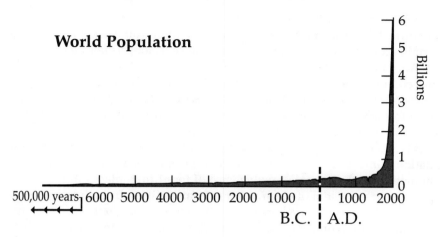

Fig. 14.4   The pattern of human population numbers over time. From *Population Bulletin* 1962.

population growth from the perspective of the ecology of humans as a unique species.

Human adaptation to the environment has involved a vast array of inventions, many of which have adjusted the relationship between population size and carrying capacity. For example, the invention of basket construction allowed humans to harvest and carry surplus food found at one place to another and to store that food safely until it was required. The simple basket became a tool that enabled the human to survive periods of inadequate food and to overcome one form of environmental control. Baskets, stone points, use of fire, invention of houses, clothes, and the social features of human life all have had this effect. The consequence is that human history is largely a history of invention applied to the control of environmental limiting factors. We have reached a point where many humans live in a mechanical cocoon, where contact with the unaltered environment is minimal. For such people, human will and desire, even capricious desire, is the measure of all things. Nothing must stand in the way of mankind! Their dream is a planet where all natural variation is banished or brought under human control and humankind looks out to the other planets to extend the human experience to the universe.

There is little reason to think that people who have been so successful in their control of the environment should reverse their historical patterns. Their success has been so profound that they have dismantled almost all the forms of social adjustment that fit human society to environment. The marketplace and individual power relationships are claimed to be sufficient to regulate human social dynamics. Reproduction and sexual behavior that were carefully controlled to maximize the likelihood of surviving offspring have become a form of private and public amusement. Death, the termination of human relationship, is delayed as long as possible.

Although the advantages of invention are concentrated in a few classes and countries, advocates of the modern life have been so successful in advertising the attractions of their life style that people almost everywhere want to share in this humanized world. The desire to be free of the unpredictable impacts of carrying capacity is universal.

My interpretation of the consequences of this history is as follows. Population numbers are important because they are basic to any process. It is immaterial if the process involves use of fossil fuel, places to park automobiles, or provision of clean water to communities. In each case the calculations begin with numbers of people to be served. So numbers are important and the population curve signals that they are increasing rapidly. Indeed, it is a reasonable guess that the human population of 5 billion will double in about forty years. The United Nations, however, predicts that the population will level off at 7.5 billion in about the same time period. In either case, numbers of humans will increase. The question is, can we provide continued growth in material inventions so that a larger proportion of humans share in modern life and are freed of their exposure to environmental carrying capacity?

The problem involves a balance between inventiveness and social regulation of the use of invention for human welfare. Invention is almost always an output of the individual human mind located in a situation that is supportive. In that case, the "ahaa" becomes useful. Rigid, highly controlled social systems seldom generate inventions benefiting society. Rather, individuals focus on inventing ways to cope with and overcome rigid controls. Invention continues, but it does not pay off for society; it becomes an opponent of society and results in conflict. Nevertheless, even in a hospitable environment invention can be highly disruptive. It can improperly benefit those who have exclusive access to it, it can exploit members of the society by changing the social contract, and it can destroy the environment. For these reasons invention must be encouraged but also managed. Few societies are able to manage this contradiction. Instead we oscillate between excessive control and freedom, paying the costs associated with each extreme as the pendulum swings from one side to the other. I think that it is normal for populations to act in this way and that we should expect this pattern to continue.

Have humans through invention freed themselves from environmental control? This is a more difficult question. By it, I mean to ask whether the forces external to the human population in its modern form are ineffective in significantly altering human demography. Physical forces such as plate tectonics, violent storms, flooding, and so on will continue to affect humans living near the events, but the effects on overall births and deaths will be minimal. Rather, factors internal to the ecological system become likely control agents. These might include disease organisms associated with food-production systems, waste-processing systems, hospitals, and so on. They might include violence, especially between competitor societies, as well as the inability of bureaucracies to manage complex problems and bring resources to problems at the proper scale. The most difficult challenge may be to recreate social controls after we have dismantled most of the traditional systems that had evolved to meet environment and social problems. At present humans are especially vulnerable in this regard.

If modern human society fails to meet the challenges created by numbers of individuals and their material demands, there will probably be a population decline and a readjustment to the carrying capacity. This will be a new pattern and will not recreate some past human experience. Humans are unlikely to go extinct. But if humans do go extinct, then the surviving biota will evolve to repopulate the planet and the ecological processes will continue until the sun stops providing solar energy.

This analysis is derived from the concept of population. It does not take account of personal ethical perspectives about the human condition. I will treat that point of view elsewhere.

Readings

Braudel, Fernand. 1979. "Weight of Numbers." Section 1 of *The Structure of Everyday Life*, vol. New York, Harper and Row.

Cohen, Joel E. 1995. *How Many People Can the World Support?* New York, Norton.

Chitty, Dennis. 1996. *Do Lemmings Commit Suicide?* New York, Oxford University Press.

Ehrlich, Paul R., and Anne H. Ehrlich. 1970. *Population, Resources, Environment: Issues in Human Ecology*. San Francisco, Freeman.

Kingsland, Sharon E. 1985. *Modeling Nature: Episodes in the History of Population Ecology*. Chicago, University of Chicago Press.

15

Life-History Adaptations

The life history of an organism involves the processes of birth, growth, reproduction, development, and death. It includes those processes that underlie increase and decrease of the population, discussed in the previous chapter. I have noted that the interaction of an organism or a population with its environment produces a pattern of responses that result in the organism's adaptation and survival or demise. These patterns have a genetic basis, because the organism's capacity to respond to environmental factors is inherited.

It is important that we take a dynamic point of view of adaptation. The organism is not a mechanical robot that reacts to an external physical-chemical environment following a set of physical laws. Living organisms are in a dynamic, interactive relationship with their environment. Even plants that do not move as individuals still produce root exudates, leaf litter, and chemical flows from stems and canopy which create a more suitable environment for the individual plant and its offspring. Higher animals are much more active. The beaver, for example, alters the hydrologic cycle, forms new landscape patterns, and has an impact on vegetation.

Nevertheless, the ecologist understands that these interactions are the result of the natural selection of genetically grounded traits. The organism has no plan that guides its choice among options, with the intent of maximizing its control of resources or improving the efficiency of its performance. On this point the philosopher-humanist and the ecologist-biologist may differ. This problem arose earlier when we considered Lovelock's Gaia Hypothesis. The ecologist is able to explain adaptation from our knowledge of genetics, behavior, and physiology. There is no need to resort to a teleological explanation that attributes purpose to organisms, populations, and communities. Alternatively, the philosophically minded person may find purpose in the natural world.

Sometimes, however, ecologists have made this issue more complex than it need be by their choice of language. For example, when we connect certain traits with certain outcomes, we speak about tactics or strategies as if organisms have options among which they choose to enhance their survival and reproduction. The logical foundation of this terminology is that organisms are given a purpose—to survive and reproduce—by the human observer, and that nature selects the most fit adaptive patterns—or strategies—of organisms for survival. Thus, it is not the organism that has the strategy, as we would employ the term in human planning; rather, it is evolution as a process that results in successful or unsuccessful strategies or adaptations. I have used the word *adaptation* in place of strategy to avoid this ambiguity.

Life History

Life-history traits include any trait under selective pressure that may influence the survival and growth of the population. In life-history studies (Stearns 1976), theoreticians and experimentalists initially examined factors influencing the rate of population increase, including the sex ratio, reproductive effort for different age classes of females, age at first reproduction, and age structure of the population. Research focused on the role of single traits on population growth; the more recent approach has been to study the interaction of groups of traits. The early studies led to the identification of several key questions, including the number of times an organism should reproduce in its lifetime and the significance of variation in clutch size, age at first reproduction, and the size of the young at birth.

David Lack, an English ornithologist at Oxford, was one of the early investigators of life-history strategies of birds. Lack (1947–1948) observed that birds in the tropics tended to have smaller clutches of eggs in the nest than did birds in temperate and arctic regions. This observation seems counterintuitive, because, as I have noted, tropical forests have higher net primary productivity and greater diversity than temperate forest and arctic communities. One might think that the tropical conditions would allow birds to have a larger clutch size under richer food conditions. Lack's explanation of this observation was that birds tend to have a clutch size which fits the number of offspring the adults can feed successfully to the point where they leave the nest. The more northerly environments tended to have longer day length than tropical environments during the fledging period and the parents could feed for more hours of the day, obtaining more food for larger broods (table 15.1). As shown in the table, birds in the arctic environment had slightly larger clutches than did British birds. Birds in the tropics had the smallest clutches.

Table 15.1 Difference in clutch size in birds from a range of environments compared to those from Britain. A negative value indicates that the average clutch size in that habitat was less than the clutch size in the same group of birds in Britain. Data from Lack 1947–1948.

Bird Group	Arctic Norway	Hot/Dry Southern Spain	Tropical Island Canaries	Tropical Mainland India
Passerines	+0.4	−0.4	−1.8	−1.3
Owls & Falcons	+0.4	−0.6	−1.3	−0.2

Clutch size is an important life-history trait in bird populations. Work carried out since Lack's studies in the 1940s (Stearns 1976) has produced several hypotheses to explain differences in clutch size. Each emphasizes one or another aspect of the biology and ecology of the species and illustrates the difficulty of arriving at a satisfactory answer to the question of adaptation. For example, one hypothesis states that an organism may have as many young as it is physiologically capable of producing, or, alternatively, that birds lay as many eggs as they can physically cover in the nest. This hypothesis focuses on the size and physiology of the bird. The organs that produce and lay the eggs are limited in size and capacity and the wing span and body size of the bird are limited, so that these anatomical features provide a limit to the number of eggs laid or covered in the nest. The anatomical and physiological explanation places the foundation of the adaptation in the history of the organism, when the size and form of the bird were formed under past selective forces.

A second set of hypotheses focus on the interaction of the bird and the environment. As Lack proposed, the parents produce, on average, the clutch size that results in the most young surviving to maturity. The clutch size is controlled by the amount of food the parents can gather, which is influenced by natural rates of plant or insect productivity. Parents that produce clutches which are too large become exhausted hunting for food and feeding young; even so, the amount of food overall may be inadequate for all the offspring. At the other extreme, parents who lay clutches that are too small can feed and fledge the clutch, but their genes contribute a smaller amount to the gene pool of the population. Eventually, their traits may become insignificant. Both extremes are selected against. The best-fit traits are those which fit the food to offspring, allowing as many to survive as possible and thereby permitting their genes to be a larger proportion of the gene pool of the next generation.

A distinction should be made between short-term regulation of populations and long-term evolution of reproductive strategies. In the short term, birds probably reproduce as rapidly as possible within their genetically controlled limits. Population size is then limited by density-dependent mortality based on physiological, behavioral, and ecological limits of the genetic potential. Over the long term, natural selection favors traits that allow the individuals possessing them to maximize their overall genetic contribution to subsequent breeding populations. Theoretically, there are several ways that this can be done. The r-selection strategy (r refers to the intrinsic rate of increase discussed in the preceding chapter) involves an early age at first reproduction, large clutch size, a single burst of production of young, no parental care, small numerous offspring, and a short generation time. The K-selection strategy (K refers to the asymptote of the population curve discussed in the previous chapter) involves delayed reproduction, repeated production of young, small clutches, parental care, and fewer but larger offspring. These two general strategies are appropriate for different sets of environmental conditions.

Life-history strategies of plants, as well as animals, have been studied by ecologists. In the case of plants, biomass is frequently used as the property of interest rather than population density. Biomass is tied to the processes of production and decomposition (see Chapter 12). For example, J. P. Grime (1977) classifies the factors limiting plant biomass in any habitat into two categories. The first contains factors restricting production, the second those causing destruction of the biomass. He calls the first stress and the second disturbance. If one divides each of these conditions into low- and high-intensity and then contrasts stress and disturbance in a four-cell matrix, one has a set of environmental conditions to which plants must adapt. Actually, there are only three conditions because the high-disturbance, high-stress case will not support plant growth at all. Grime related plant adaptive types to the three environmental situations as follows: (1) low-stress and low-disturbance environments select for competitive plants, (2) high stress and low disturbance select for stress-tolerant plants, and (3) low-stress and high-disturbance select for ruderal or weedy plants.

Plants in these three different conditions of natural selection have the following characteristics. The first involves selection for highly competitive activity which maximizes vegetative growth in productive, relatively undisturbed conditions. The second involves endurance of continuous unproductive conditions due to stress and/or resource depletion through reduced vigor. The third is associated with a short life span and high seed production. Each of these strategies results in a set of characteristic traits which are expressed morphologically, physiologically, and ecologically in different life forms (fig. 15.1). For example, the leaf form under the three strategies varies from a robust shape in the competitive strategy, to a small, leathery, sometimes needle-shaped form under the stress strategy, to several

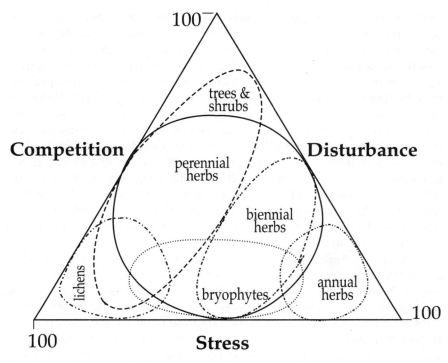

Fig. 15.1 A model, modified from Grime by MacMahon (1980), depicting equilibria between competition, stress, and disturbance in vegetation. The three axes are the percentage of each strategy. Approximate positions of plant life forms are indicated.

different forms under the ruderal strategy. This pattern is observed in the dominant oak trees of dry, sandy, nutrient-poor, stressful habitats in the southern United States. These oaks tend to have small, evergreen, leathery leaves that contain large quantities of tanins which restrict herbivory. In contrast, on nutrient-rich, moist sites the oaks are of different species that have larger, broad-shaped leaves.

Stephen Stearns in his review (1976) and H. Caswell in a more recent synthesis of life-history strategy research (1989) point out that the fundamental premise of life-history theory is that these traits are part of the expressed genetic feature of the organism, and therefore are as subject to genetic explanation as are traits of morphology, physiology, or behavior. The task of the life-history researcher is to predict which traits will evolve in which organisms living under what conditions. Stearns summarizes some of these predictions:

1. Where adult mortality exceeds juvenile mortality, the organism should reproduce once in its lifetime.

2. Brood size should maximize the number of young surviving to maturity, summed over the life of the parents. But when optimum brood size varies over time, smaller broods are favored because they decrease the chance of total failure of reproduction.

3. In expanding populations, selection should minimize age at maturity. In stable populations, reproductive success depends upon size, age, and social status.

4. Young should increase in size at birth with increased risk of predation and decrease in size with increased resource availability.

Implications

The concept of life-history adaptations is a view from both the organism and the population sides of the ecological whole. It focuses on the prime element of the organism's life, which is its capacity to adapt and evolve. The consequence of the processes is species diversity, ecological complexity, and survival of life on the planet. Although it is simple to express these ideas in language, it is much more difficult to express them in quantitative terms and to tease apart cause and effect.

This problem has several parts and they raise important questions about our capacity to understand our environment. The first part of the problem concerns the variation in and selection of life-history traits. These traits, such as clutch size in birds, must vary from one individual to another in order for selection to occur. Yet clutch size is itself a complex of morphological, physiological, and behavioral factors, each of which is under genetic control and is involved with other life-history traits of the species. We represent this twisted set of interactions by the word *complex.* It is complex, indeed, and it is the subject of the science of genetics and developmental biology.

The second part of the problem is that the environment is multifactorial and variable. The environment is also a complex. So in the analysis of life history we have one complex approaching another complex! This is a tremendous challenge to the scientist, and the search is on for specific cases or mechanisms that would reveal how the complex might work.

Another part of the problem is the difficulty of carrying through experimental work on complex, multifactorial problems. The experimental approach usually requires the manipulation of cause and the following of effect, in a controlled environment. The result is a statement that a specific relationship between cause and effect was observed under specified environmental conditions. If the experiment has been properly designed, the investigator is able to state the likelihood of the relationship being observed in other experiments under similar conditions. If the model is made more complicated by adding a second causative factor, the investigator must show the effect of cause 1, the effect of cause 2, and the effect of the interaction of causes 1 and 2. The solution is not merely additive; interaction is an important additional element in the analysis.

This brief discussion suggests the nature of the problem. Most environmental factors (cause in the cause-and-effect equation described above) are multiple and ever-changing, so that the analysis deals with many factors, which have different levels of significance in causing an effect on the endpoints of interest. In such a case we move beyond the capacity of the experimental approach to untangle cause and effect. Although some of the experimentalists, such as the Illinois ecologists E. A. Desy and George Batzli (1989), have carried out three- and four-factor replicated field experiments, such studies are relatively uncommon. Rather, we must move to a modeling approach, where we construct a simulation of the complex pattern on a computer and study the effect of varying inputs on outputs of the model. In this abstract way we can learn how complex systems might function in the unconstrained natural environment.

The difficulty of analysis of complex patterns such as life history leads us away, emotionally and psychologically, from the arrogance of the person who operates in a designed and mechanical environment. I have discussed the implications of a world where the environment is created by humans and is therefore "known" in a fundamental sense, compared with the natural world which is "unknown" or only partially known. Even a very complex machine like a computer can be made to operate within defined limits.

The natural world is not a machine. Its complexity requires that those who live in and work with it have the patience to accept the rate at which natural processes operate in time, the openness of mind to accommodate variance and fluctuation, and the humbleness to recognize when we do not understand.

Human managers frequently try to manipulate the behavior of species in the environment. For example, we attempt to destroy pests and predators that compete with humans. Our intolerance of these creatures is total, and we are willing to poison ourselves in order to control them. But we almost always fail because the adaptive capacity of the organisms results in a response to the human-imposed selective force. Control works for a while and then there is a resurgence of the pest. It is folly to try to limit this capacity for change. A more ecological approach to the problem is to work with these natural processes, directing them strategically so that an accommodation between humans and organisms is achieved. Of course, compromise means that humans cannot wring every bit of energy and material out of the ecosystem for ourselves. But it also does not mean that we have to give up agriculture, forestry, hunting, and fishing and all the other uses we make of the natural environment.

Readings

Desy, E. A., and G. O. Batzli. 1989. "Effects of Food and Predation on Density of Prairie Voles: A Field Experiment." *Ecology* 70: 411–21.

Emlen, S. T., and L. W. Oring. 1977. "Ecology, Sexual Selection and the Evolution of Mating Systems." *Science* 197: 213–23.

McNab, Brian. 1980. "Food Habits, Energetics and the Population Biology of Mammals." *American Naturalist* 116: 106–24.

Tamarin, R. H. 1978. "Dispersal, Population Regulation and K Selection in Field Mice." *American Naturalist* 112: 545–55.

Willson, M. F. 1983. *Plant Population Ecology.* New York, Wiley.

16

The Individual Organism

The individual organism is of fundamental interest to ecologists. We observe and measure the individual organism in the field. Our theory is based ultimately on the individual. The individual is the basic element of the groups that make up the other levels of ecological hierarchies. Yet for many organisms we have considerable difficulty determining what is or is not an individual. For example, a rhizome of Bermuda grass (*Cynodon dactylon*) grows at the soil surface and puts up shoots along its length. These shoots appear to be individual plants growing from the soil surface and we may count these shoots as individuals, yet they are all connected by a network of roots and shoots. An aspen forest may be interconnected through the roots, so that the entire forest is a single individual plant (Mitton and Grant 1996). A coral is even more complex because in the coral head an algae and an animal combine to form what we see as an individual coral polyp.

Furthermore, an individual is arbitrarily separated from its environment by our method of analysis. From this perspective, we see an ecological system in which the individual organism and the environment interact as distinct entities. Ecological research has tended to focus on the individual in this context. A different perspective, in which the individual is perceived as a system where the organism and environment are merged, is less attractive to researchers because it avoids the use of dualism or cause-and-effect logic. It leads to description of the whole, not mechanistic explanation of the parts. Another perspective is to take a social point of view and define the environment of the individual as other organisms which interact with the individual. The consequences of our definitions and approaches are not trivial.

The individual ecological system has all the properties that I have described for the other types of systems. It has a biomass, contains energy and mineral elements, processes and cycles these materials, adapts and changes, and has a life history. The unique aspect of the individual, and the reason why it is so central to ecological studies, is that it is the locus of biotic creativity through which evolution and natural selection operate. It is probably fair to say that the great divide in ecological science is between ecologists who base their studies on the laws of thermodynamics and conservation of mass (as described in Cluster 2) and those who base their approach on the theory of evolution (Clusters 3 and 4).

The remaining chapters of Cluster 3 will examine various attributes of the individual organism. We will begin, in this chapter, by treating the individual as an ecological system that has an energy flow. This perspective links our examination of the individual to the other systems we have considered in earlier chapters and with the subject of plant and animal nutrition, metabolism, and growth. In this chapter the emphasis will be on how food energy is managed in the organism. In the next chapter we will examine the physical energy environment of the organism. Then we will consider evolution and natural selection.

Energy Flow Within Individual Organisms

The energy-flow diagram for the individual animal can be used to identify the important energy fluxes that we should consider (fig. 16.1). The intake of energy is primarily through feeding (although the body temperature of the organism is influenced by conductance of heat from the environment, especially in those organisms that do not regulate their body temperature physiologically). Not all of the food energy consumed can be digested and assimilated by an organism (table 16.1). Meat eaters digest and assimilate a larger percentage of consumed energy than do herbivores. Only a part of the food can be reduced chemically to a state where it can pass through the walls of the intestinal tract and enter the bloodstream. This first divide in the energy flow through the organism reduces the quantity available to the organism by shunting part of the intake energy to the waste stream. For a mammal such as a deer or a fox, the quantity of energy assimilated is usually determined by subtracting the energy excreted in defecation and urination (the waste streams) from that consumed.

The energy that is assimilated is available for metabolism and for building tissue. Metabolic activity requires the greatest amount of energy flow. Considering mammals and birds, each individual has a basic metabolic requirement. It is called the basal metabolic rate and represents the energy cost when the organism is at rest and not digesting food. The basal rate varies depending on the temperature of the animal's environment (fig. 16.2). As the temperature declines, the animal usually increases its metabolism to maintain body functions, although some species can use behavior

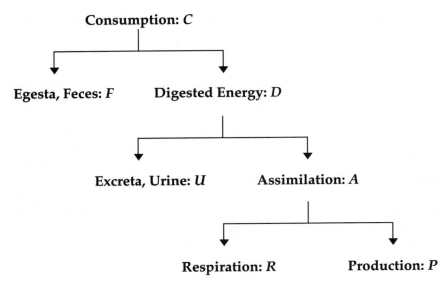

Fig. 16.1   Energy flow through a mammal or a mammal population.

and insulation to resist energy loss. The arctic fox (*Alopex lagopus*), for example, can cope with the coldest temperatures it might encounter in its harsh environment without greatly increasing its metabolism because it has an insulating fur coat and a habit of curling into a ball to protect its nose and mouth (Scholander et al. 1950).

Similarly, as the temperature increases, the metabolism rises and eventually the animal dies from heat stress. Animals seem to have less capacity to regulate metabolism on the hot, as compared to the cold, end of the temperature gradient, and as a consequence the metabolism-temperature curve is much steeper. There is a point between these extremes where metabolism is minimal, called the thermoneutral zone. The basal metabolic rate is properly measured at this point.

Table 16.1   Digestibility and assimilation of natural foods, as a percentage of the energy or organic matter consumed. Data from Grodzinski and Wunder 1975

Feeding Type	Number of Species	Digestibility (%)	Assimilation (%)
Grazing herbivore	14	67	65
Omnivore	10	77	75
Granivore	6	90	88
Insectivore	9	——	85
Carnivore	2	90	——

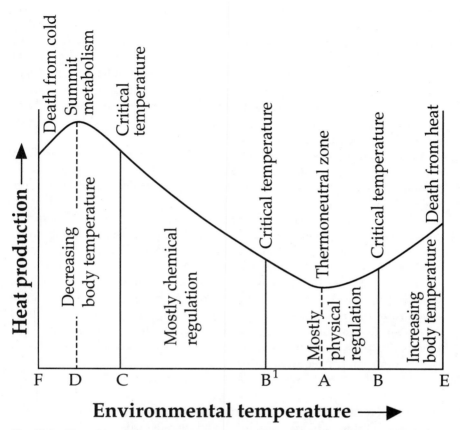

Fig. 16.2 The influence of temperature on energy production. Redrawn from Brody 1945.

As the organism goes through its daily activities, it expends additional energy above the basal level. The organism's ability to do this is limited by the need to maintain a positive energy account. The organism cannot expend more energy than it obtains, except for short periods of time when it uses energy stored in body fat and muscle.

Each activity has its own energy cost. For many mammals the energy required for daily living is about double the basal metabolic energy requirement. Certain activities, however, greatly exceed this amount. For example, a mammal at peak work can expend seven or eight times its basal rate for sustained periods and twenty times basal for short spurts, as when a predator chases its prey.

An important energy and material requirement of the individual is to grow, mature, and then reproduce. If energy is limited, these primary activities may not be carried out. For example, a female snake consumes other animals and grows in size (Congdon et al. 1982). At a certain size, growth stops and excess energy is shunted to fat bodies in the female. When the fat bodies have reached an appropriate size, accumulation of fat stops and eggs are developed. Once the eggs are laid the process starts again. The point is that the individual must satisfy its personal energy requirement daily or regularly, but it also must provide for the next generation. This additional demand must be satisfied for the species to survive.

In order to balance the energy account, the organism must find food containing energy to fit its requirements. Foods differ widely in their energy content. For example, leaves and stems have a relatively low caloric value for herbivores—around 4.5 kilogram calories per gram of dry weight. In contrast, many seeds have a value of 5.5–6.0 kilogram calories. The difference between leaves and seeds is due to the high-energy-value lipids stored in seeds. Meat has a caloric value similar to seeds, while the energy content of animal fat may be 9.5 kilogram calories.

Because of these differences in food energy content, organisms have a variety of options for satisfying their caloric requirements, and their strategy of resource use will reflect these options. For example, Jim Karr (1975) compared feeding in bird faunas from Illinois with those in Panama and Liberia (fig. 16.3). Karr speculated that tropical species had more options to satisfy their energy requirements because the tropical forest contained a greater variety of foods over longer periods of time. For example, fruits are not present in temperate forests except at very limited periods during the summer. Karr's predictions were confirmed: the tropical faunas used a greater array of food items than did the temperate birds.

These considerations of energy flow in the individual organism lead to several key points that bear emphasis. First, the energy requirement for system maintenance (the basal rate of the deer or fox) sets a demand on the organism that must be met, almost on a daily basis. Although organisms can store energy in or outside of the body or reduce their metabolic costs, ultimately an energy debt has to be paid. This requirement shapes the biology of the organism in a fundamental way. The adaptations for food processing, the ruminant stomach, the elaborate behaviors of predators, and so on are all tied to these requirements.

Second, the energy resource is in the organism's environment and the organism must meet its energy debt by collecting sufficient energy daily. The interaction between the energy-hunting organism and that part of the environment it considers a potential resource is fundamental to the ecology

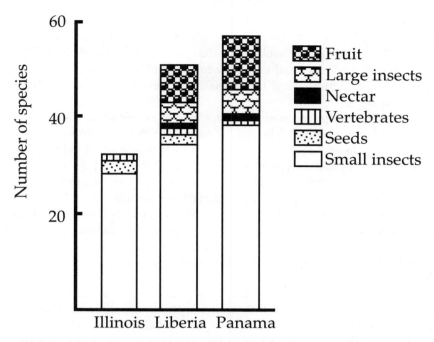

Fig. 16.3 Number of resident bird species exploiting food resources in three forest study areas, representing temperate and tropical environments. From Karr 1975.

of the individual. Many of the adaptations the ecologist notes are involved with feeding and energy collection.

These comments have almost entirely referred to mobile animals. Plants and sessile animals must meet their energy requirements too. The plants achieve this through photosynthesis and transpiration. Sessile organisms have evolved a variety of devices to attract and collect food resources that pass by them or come to them.

Implications

We are all familiar with the concept of the individual. Indeed, in our individualistic modern society we tend to give the individual such weight that all other forms of organized life are unimportant in comparison. If we view the individual from the perspective of ecology, however, we recognize that it is a living system interacting with its environment. Individuals interpenetrate and are interpenetrated by their environment. The boundary of the individual is as arbitrary as the boundary of the lake or the population. Individuals are in a constant process of physical and chemical exchange with other individuals and with the inorganic environment. The concept of the individual is as loose and as arbitrary as that for other ecological units. The significance of the individual is, first, a function of our sense perception and, second, a function of the interpretation of that sensory information through our culture.

These relationships are made clearer when we consider the energy environment of an individual. The individual must balance its energy equation. This is a life-and-death matter; there is little toleration of debt in the natural world. Animals and plants expend most of the energy they receive to sustain their lives; there is little left for storage, reproduction, or unusual activity. Thus, organisms are in a balancing act, maintaining a thermodynamically unstable situation by constant input and output of energy to and from other organisms and the physical environment. The individual maintains its dynamic structure for a period of time, set by its life history, and then its parts become reconstituted into other organisms in the ecological system.

The individual plays a special role in ethical theory, because the traditional Western view has been that human individuals alone have moral status, as a consequence of our having souls, rationality, and a separate creation. It is widely accepted that an individual human has inherent or intrinsic value even when that individual is incapable of expressing human activity. For example, our laws and moral codes protect and support the accident victim or the congenitally handicapped individual who cannot act on his or her own behalf. The intrinsic value of these human individuals is accepted in law and codes of behavior. Indeed, it is said that the way in which a society cares for such individuals is a criterion of civilized behavior.

Further, in American culture individuals have special legal rights that are guaranteed under the Constitution. The natural world has been transformed into the property of individuals by virtue of their labor in converting nature into a farm, a cut-over forest, or a trout stream. The Western ethical system is expressed in a legal system that supports the moral considerability of the individual human within the society.

Environmental ethicists seeking ways to express the value of nature have expanded this human-centered perspective to encompass organisms in nature. Peter Singer (1990) used sentience, or sense perception, and the ability to suffer pain as a justification for moral consideration. Singer argues that if a being suffers, there is no moral justification for ignoring its suffering. Tom Regan (1983), who supports animal rights, would extend moral consideration to all individuals with a desire, perception, memory, feeling of pleasure or pain, and the ability to take action to achieve goals. In his view, individuals with these properties have moral rights and must be treated in ways that do not deny this kind of value.

Other ethicists extend this form of reasoning beyond humans and domesticated animals and the "warm and fuzzy" creatures to the other elements in the natural world. For example, Paul Taylor (1986) adopts a biocentric, as compared to an anthropocentric, world view and considers that all living organisms have inherent worth. The core ideas in Taylor's argument are the interdependence of species and a denial of human dominance over nature. Taylor asks that we not harm, interfere with the freedom of, or deceive other creatures, and if we do harm them, Taylor calls for us to make restitution.

Readings

Johnson, Larry E. 1991. *A Morally Deep World: An Essay on Moral Significance and Environmental Ethics.* New York, Cambridge University Press.

Mitton, J. B., and Michael C. Grant. 1996. "Genetic Variation and the Natural History of Quaking Aspen." *Bioscience* 46, no. 1: 25–31.

Ralston, Holmes III. 1988. *Environmental Ethics: Duties to and Values in the Natural World.* Philadelphia, Temple University Press.

17

Body Size and Climate Space

Possibly the most important property of the individual from the perspective of ecological relationships is its body size. The term *size* may refer to a variety of body dimensions depending on the kind of organism and the process. Usually size is defined as biomass, because it is a relatively simple task to weigh an organism and determine its weight. Another common dimension of size is surface area. Surface area is useful because chemicals cross the surface of the organism and the area of the surface regulates the rate of flow. In plants we frequently measure leaf area and express the rate of photosynthesis per area of leaf. Mass and area are related dimensionally, so that we can move back and forth between these two measures of size.

The processes that characterize an organism, such as its flow of energy, the movement of nutritionally required chemicals, and its behavioral activity, involve interactions with the environment. The organism occupies a space, which is called a home range, territory, or habitat, where it finds the resources it requires. There is a physical space for each of the major resources in which that resource is available to the organism in adequate quantities to support life. Ecologists concerned with the temperature conditions of the environment as they affect the organism have referred to the area with desirable temperatures as a "climate space." Collectively, all of the spaces representing separate limiting environmental factors form the home range or habitat space.

To explore the concepts of body size and climate space we will continue our focus on energy flow in the individual. But in this case we will enlarge our view of energy to include that which interacts with the organism through conductance, reflectance, and convection from the energy environment.

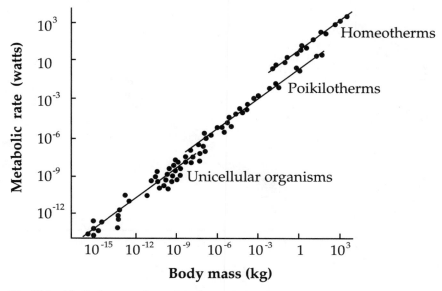

Fig. 17.1 Metabolic rates of unicellular organisms, poikilotherms (animals that are unable to regulate their body temperature), and homeotherms (animals that can regulate their body temperature). Metabolism is expressed in watts; one watt equals 0.01433 Kcal per minute. From Peters 1983.

The Role of Body Size

Body size affects the energy flow of the individual organism through the intake and output of energy across the organism's surface. According to the physical laws of Newton and Stefan-Boltzmann, the rate of cooling of a body is proportional to its surface area. Since heat loss and heat gain must be balanced, heat production is also proportional to the surface area. R. H. Peters (1983) explored these relationships, and his analyses have been used to fashion this account of energetics.

Surface area is difficult to measure directly. Empirical studies have shown that the surface area of an organism is related to its body weight and we can use this relationship to estimate metabolism. Metabolic rate is related to two-thirds of an organism's body weight across a wide variety of organisms (fig. 17.1). Animals that expend energy to regulate their body temperature have a higher overall relationship than those organisms that do not regulate their body temperature, but the slope of the relationship is the same for both groups.

Surface area per unit of body weight and energy metabolism per unit of body weight decline with increasing body weight. As an organism increases in size, its weight increases at a faster rate than does its surface area.

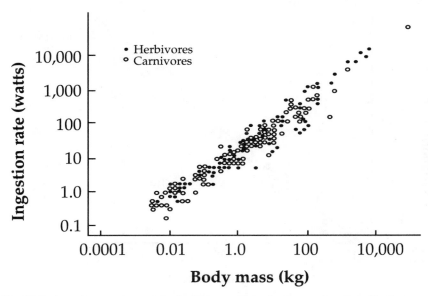

Fig. 17.2 Ingestion rate in watts for herbivores and carnivores, as a function of body size. From Peters 1983.

For this reason a large animal has a smaller surface area in relation to its mass than does a small animal. Because energy flow is regulated by surface area, a large animal requires less energy per unit of weight to maintain itself than does a small animal. This relationship is one of the reasons why animals can be only so small. For example, the shrew is one of the smallest mammals. It has to be very active, because it must eat frequently to obtain sufficient energy to sustain itself. Its metabolism per gram of body weight is many times greater than that of a larger carnivore, such as a lion.

Since the largest proportion of energy taken in by an organism is employed in maintenance, it is reasonable to predict that food intake would also be related to body size by the two-thirds factor. It turns out that this is true for both herbivores and carnivores, across a wide spread of species (fig. 17.2).

Because animals move about on the Earth's surface collecting energy at speeds and distances equivalent to their energy requirements, the home range size (that is, the area over which they hunt) also is related to body size in the same way as metabolism, although the relationship is not nearly as tight (fig. 17.3).

Finally, if the area an animal lives in, its food consumption, and its metabolism are all interrelated, then the numbers of animals also may be related (fig. 17.4). This prediction is also realized (Peters 1983). The density

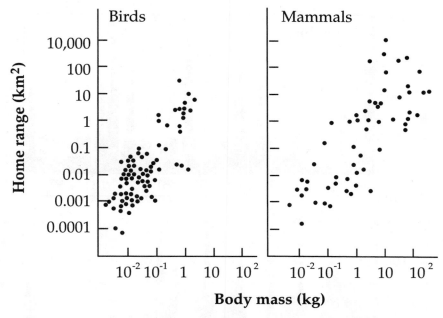

Fig. 17.3 Relation between home range and body size in birds and mammals. From Peters 1983.

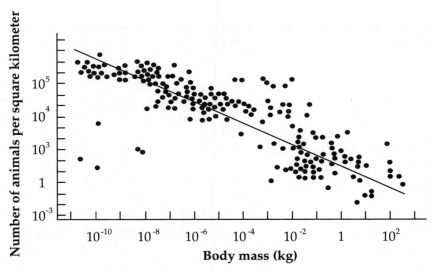

Fig. 17.4 Relation between body size and population density for mammals, birds, invertebrates, and herpetiles. From Peters 1983.

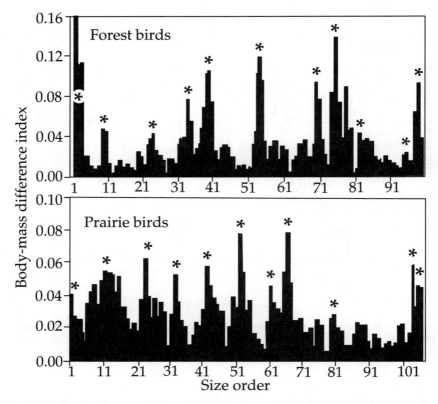

Fig. 17.5 The distribution of body size of birds in boreal forest and in the North American prairie. From Hollings 1992. The abscissa represents the difference in body size, so that gaps appear as peaks in the figure. Gaps are also indicated by asterisks.

of birds, mammals, reptiles, and invertebrates all follow a linear negative relationship with body size. The relationship suggests that large animals will have lower densities than small animals—which is what we observe.

C. S. Hollings (1992) takes a somewhat different approach to body size. He examined the gradient of size within landscapes, in contrast to considering all animals within taxonomic classes, as Peters did. Hollings noted that the distribution of sizes was discontinuous, with an absence or reduction of organisms of certain body sizes (fig. 17.5). That is, Hollings noted "lumps" of organisms of a given range of body size within specific landscapes and observed that these patterns differed by landscape. He postulated that these discontinuities indicate the scales of resource properties of a landscape and the hierarchy of animals living in that landscape. The interaction between these two hierarchies produced the observed discontinuities. This is an especially interesting example of the linkages of organism and environment, through hierarchies of scale.

Climate Space

The energy flow of organisms is not fully explained by the intake of energy in food and the use of energy in metabolism and reproduction. Organisms also are living in an energy environment that affects the organism energy dynamics in ways other than as packets of food. The ecologists Wayne Porter and David Gates developed a concept called climate space to describe the energy environment of the organism. By climate space they mean that there are energy environments within the ecosystem where an organism with a particular metabolic rate, food habitats, activity pattern, and body insulation can live. There are also energy environments where the organism cannot live, because they cause too rapid an energy loss or gain that cannot be compensated for by metabolic, behavioral, or insulation adaptations. If the organism enters these unfavorable environments, it loses or gains heat and dies. Thus, the climate space defines where an organism can survive. Can we predict the dimensions of this space?

The question can be explored by looking at the experience of Porter and Gates with the desert iguana (*Dipsosaurus dorsalis*) at Palm Springs, California. The investigators took an ingenious laboratory approach to the problem. They constructed model organisms of the same shape and dimensions as real iguanas and measured the flows of electrical energy across the surface of the models to determine inputs and losses on various pathways and under different environmental conditions. In this way they constructed an experimental space with the dimensions fitting the model iguana's capacity to maintain its body temperature within thermal limits. They then were able to use these data to interpret their observations of live iguanas in the desert environment.

The iguana lives in an environment where the surface temperature may exceed 40 degrees centigrade (fig. 17.6). Because the iguana cannot tolerate such high heat, it spends much of its time in a burrow and radiates energy to the burrow walls, which are much cooler than the soil surface. Its activity on the surface is restricted to short periods in the morning and evening when the temperature is cooler (fig. 17.7). It is also influenced by the presence of shade from shrubs and by air movement, both of which reduce the direct energy input to the iguana. By such behavioral adaptations the lizard stays within its climate space (fig. 17.8). By controlling its metabolism and energy flow, the lizard can build up the fat reserves required for reproduction. These behavioral adaptations enable the species to survive in an exceptionally harsh environment.

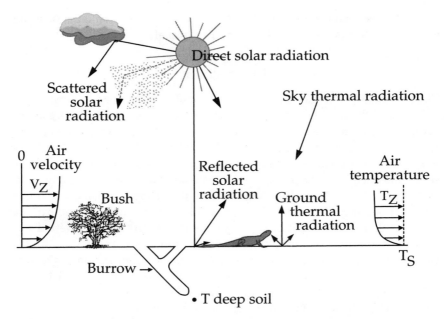

Fig. 17.6 Schematic diagram of the desert environment, showing energy flows to and from the surface and to a lizard. From Porter et al. 1973.

This is an extreme case, but it illustrates the principle well. The environment provides both the food required by the organism to meet its energy debt and a variety of climate spaces in which the organism can find suitable temperature conditions. Both of these energy requirements must be met for survival and reproduction.

Implications

Size is important in nature and it is an important predictor of the life form of the organism. It is also fundamental in human activities, although this property is not widely understood. Indeed, in the American culture, bigger is frequently deemed to be better! Contrary to the slogan, size controls process, with the result that there is an optimum size for most, if not all, processes. This is true of the body size of humans, and it is also true of social groups, classes, universities, cities, and so on. The human-built environment is subject to the same physical laws as the natural environment. As the size increases and exceeds the optimum, special efforts are needed to maintain the direction and purpose of the system. Management of this

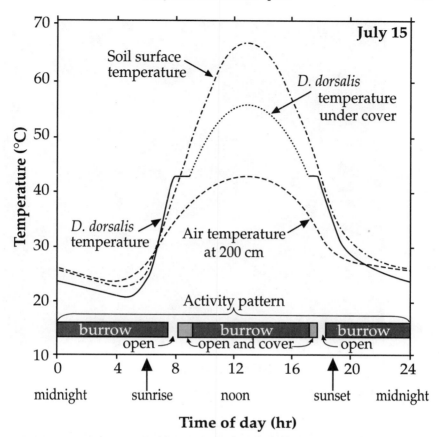

Fig. 17.7 Predicted environmental and lizard temperatures at Palm Springs, California, on July 15. From Porter et al. 1973.

problem has typically involved creating hierarchies of control. The too-large system is broken into units near the optimum size and individuals representing each subunit are grouped together to form a superunit, with control functions. This device works for a while, but it too is subject to size limitation. Ultimately, too many levels exist between the lowest units and the highest unit to maintain contact, and the higher levels are perceived as dictatorial. The result of too great a size and lack of contact can be increased maintenance costs, social breakdown, violence, and personal frustration.

Bigger is not always better, and small is not necessarily beautiful! Rather,

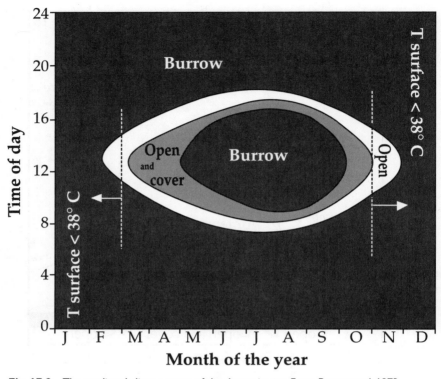

Fig. 17.8 The predicted climate space of the desert iguana. From Porter et al. 1973.

there is a relationship between size and process, with an optimum position between smallness and bigness. Going beyond this optimum in either direction generates greater and greater costs that may ultimately cause collapse of the system.

Readings

Gates, David M. 1962. *Energy Exchange in the Biosphere*. New York, Harper and Row.

Gessamen, James A. 1973. *Ecological Energetics of Homeotherm: A View Compatible with Ecological Modeling*. Logan, Utah State University Press.

18

Speciation and Natural Selection

Field ecologists are frequently impressed by the enormous numbers of organisms they encounter. A few spadefuls of soil, a meter-square quadrat in an abandoned field, the contents of a plankton net pulled through the surface waters of a pond reveal a richness of biological life that often overwhelms the ecologist's taxonomic skills. It is commonplace under such circumstances to find species that are new to science; that must be studied and described by a specialist before the ecologist can be confident of their name. As discussed in Chapter 11, this richness is called species diversity, and it is an important characteristic of communities, indicating the nature of their isolation and size as well as the richness or poverty of their environment. When the ecologist focuses on a single population, he or she encounters a similar richness of individual organisms, which vary in their size, physiological responses, behaviors, and the range they occupy. Everywhere we look we find biological variety. Our observation of biological richness raises the question of its origin and maintenance. How is such variety generated?

To answer, it is necessary to enter the domain of genetics and cell biology. Our question is a basic one within biology and, in one sense, all of biology is an examination of differences between organisms, organ systems, and cells in species of microorganisms, plants, and animals. An answer also is of central importance to ecology. Indeed, it is so important that ecologists have combined with geneticists to form an interdisciplinary subject called evolutionary ecology. Many modern ecology texts begin with evolution and then consider how individuals are organized into populations and communities. In these books the systems approach is left for the end, if it is included at all.

Mechanisms

Biological variety has two sources; one is within the genetic structure of the cell and the other is in the environment. The interaction of these two dynamic sources produces the variety we observe in the field.

The genome. The cell contains a genome, which consists of chromosomes on which are coded the instructions that guide cell division, development, and the basic functions of the organism. We begin as a single cell and as this cell divides to produce the multicelled organism of the adult plant or animal, the genome is duplicated and transmitted to each cell in the organism.

Certain cells are specialized to engage in reproduction. In these the process of division is different than in the other cells in the body. Prior to reproduction, the paired chromosomes separate and only half of the normal complement is located in each germ cell. When the germ cells of each parent combine, the full complement of genetic material is restored in the new cell. The new cell obtains part of its genetic information from one parent and part from the other—in this sense it is new, unlike either parent. It is this process of sexual recombination that is the source of much of the biological variety we observe.

Of course, there is variety in the process itself. If the fertilized germ cell divides and forms two or more identical cells, which then continue division to form individuals, they will be identical twins, triplets, and so on. Or two germ cells could be fertilized at the same moment and both implant and grow into viable offspring. These fraternal twins would have a common birthdate but not the same genome. The variability in process is fundamental, and everywhere we look we find alternatives in structure and function. This is an example of a core principle: biological systems generate variation, but within an order that is represented by the genetic code itself, the cell, the process of reproduction, and the relation of individual organism to its environment.

Although recombination is a source of biological variety, it is not the only source. The process of cell division and development is complex, and there is abundant opportunity for rearrangement of the genetic material or for accidents in replication. These accidents are transmitted to new cells. Further, external factors can have a direct impact on the genetic system. For example, cosmic radiation may strike the chromosome, breaking it and causing change in the genetic structure. We call these changes mutations, and they are a serious result of exposure to nuclear radiation.

Thus, within the cell itself we find an orderly process of rearrangement of the genetic information and other sources of variation that are accidental, random, and unexpected. If these sources cause too extreme change, they can be detrimental to the organism and cause its death. If they are small, they may be insignificant and only appear later as they accumulate through recombination.

The environment. Thus far we have treated cells and organisms as if they were floating in ether, interacting exclusively with each other. However, cells and organisms function in an environment. The environment provides the resources and services the organism requires to live. As I have mentioned before, the environment is variable and the resources and services an organism requires are seldom abundant or homogeneously distributed in space and time.

One way to approach this problem mentally is to think of the environment as streams of flowing resources, which intermix and diverge somewhat as rivers do in their deltas. The biological cell or organism moving across these streams finds variation in the resources and services it needs to maintain life. If it can stay within a region of richness or alter the stream in some way so that richness is preserved, then the organism might prosper. If it cannot, it will not grow and reproduce; it might even die. The variety in the environment is another source of the biodiversity we observe in nature.

Natural Selection

In the formation of the ecological system consisting of organism(s) and environment, two forms of variation come together in space time. On one side the source is within the biological entity; on the other side the source of variation is in the physical-chemical history of the Earth. Both are continually changing, but the nature of change is different in that one follows biological rules and the other follows physical-chemical rules. In coming together they create ecological reality.

It is a property of the biological side that variation is produced in excess and endlessly, as long as the environment is capable of providing the resources and services to sustain it. We see this profligate production in the spawn of salmon, in the rain of tree seeds, and in the swarms of mosquitoes. We are also impressed with the genetic differences between contiguous populations. For example, R. J. Berry (1978) describes house-mouse populations living in grain ricks that differ genetically from the population living in the next rick in a harvested field. Apparently, each rick has its genetically different mouse population.

Production exceeds the capacity of the environment to support it, and death or diminished life awaits organisms that are unable to obtain their needs from the environment. In this sense, the environment selects among the potential genomes those which will produce the next generation. But this is not the kind of selection we perform in a grocery store when we pick favorite vegetables or fruits from the counters. The environment is not an active selector of organisms, as the phrase *natural selection* implies. Charles Darwin himself used the phrase as a metaphor, stating that it was "as if" natural selection functioned as does a farmer making selection among varieties. Rather, selection involves a great deal of chance—the chance of being in the right place at the right time. And it involves the intersection of two complex dynamic processes, one biological and the other environmental.

Speciation

Each individual originates from a single cell, which divides and develops, following instructions coded into its genes, into a multicellular organism. The organisms that survive to mature age reproduce and produce offspring, thereby creating a lineage that can be described by a branching tree of descendants. All the members of the tree share characteristics, but each member is also individual and different in a genetic sense. All living organisms are members of a tree of familial relationships.

Relationship does not imply cross-reproduction. Members of one branch on the phylogenetic tree usually cannot reproduce with those on another branch. We recognize this reproductive separation taxonomically. If the members of a lineage can reproduce with each other and produce viable offspring, we identify them as a species. Species represent the closest form of genetic relationship, in which all the members potentially are capable of interbreeding and exchanging genetic information.

As a lineage expands it encounters a variety of environmental conditions. Physical barriers, such as a stream, could separate members of the lineage and prevent their breeding. Over time, genetic response to the different environmental conditions or merely drift to different gene frequencies could lead to lineages having genotypic differences that prevent breeding among the separated populations. At this time these populations would be considered two species, and the phylogenetic tree describing their inheritance would show a branch or split where the two species diverged from the parent form.

Speciation takes place for many reasons beside the presence of environmental barriers to movement. Barriers to reproduction also could arise within the biological side of the relationship. These barriers to reproduction across species are also sometimes permeable. For example, P. R. Grant and B. R. Grant (1992) show that about 10 percent of bird species hybridize naturally. Hybridization may yield lower fitness because the two resulting genetic stocks may be incompatible, or it may yield an advantage because of increased genetic variety in the offspring. Grant and Grant found that hybridization between several species of Darwin's finches on an island in the Pacific Ocean did result in higher survival and breeding success. Hybridization is another important component of speciation.

The species concept has evolved as our understanding of biology has deepened. Carolus Linnaeus, the great Swedish botanist, originated the taxonomic categories we use today (Blunt 1971). The categories are organized hierarchically and form the phylogenetic tree mentioned above. Each species is a member of a genus, which is grouped into families, orders, classes, and phyla. The technical name of a species has two or more parts. The first name refers to the genus, the second to the species. For example, *Pinus taeda* is the name of the loblolly pine. *Pinus* is the genus name, *taeda* is the specific name. To determine a species' family, order, and class, it is necessary to consult a manual of taxonomy.

Even in an introductory book it is important to note that biologists do not agree about the nature of species. Alexander Rosenberg (1985), for example, identifies several different species concepts used by biologists and ecologists. The first emphasizes reproductive isolation (Mayr 1942), as I have done above. Another emphasizes the lineage of organisms evolving separately from other lineages. Yet another emphasizes the occupancy of a resource space or a range where the species occurs. The concept of species that I have used in this chapter incorporates aspects of all of these notions.

Implications

In this chapter we have examined the most fundamental property of living organisms—their capacity to continually generate new forms of life. Life, which occurs nowhere else in the universe, as far as we know, has persisted on the planet Earth for more than three billion years, even during times when there where changes in the shape of ocean and land, extensive volcanic activity, meteor impacts, and glaciation. The resiliency to cope with these changes is based on the biota's capacity to generate biological variety.

The generative capacity of living organisms is an example of a principle that is used in systems science. In a fluctuating and unpredictable environment, that system survives which has an operating range greater than that of a potential environmental control. With this capacity, the impact of the environment may be devastating to the system, but it cannot be fatal. As described in Chapter 7, Ross Ashby, the cyberneticist, calls this the Law of Requisite Variety. Ashby applies his law to a mechanical system where system control is the focus of interest. In this case the controller (a thermostat or governor) must have a capacity range greater than the expected performance of the overall system. Otherwise, the system will not be stable. In the real world survival may be the result of the capacity to generate variety, the heterogeneity of environment, or chance. Even great cataclysms, such as the movement of tectonic plates, volcanism, or the formation of glaciers, are usually not global in their impact.

An alternative explanation for biological diversity and the fitting of biological function to the environment is design. In the eighteenth and nineteenth centuries the evidence of scientists supported the idea that there was a designer of the world. It was thought that an orderly world where birds have wings in order to fly and fish have fins in order to swim could only come from the mind of a designer, who created the rules by which the world changes. In this view, all change is anticipated and organisms adapt to the expected changes as they occur. The modern view, derived from the studies of genetics, biology, and ecology, is fundamentally different.

The implication of these concepts of genetics and ecology is that scientific knowledge is congruent with and brings us back to an earlier, less abstract form of explanation of the origin of the world and of human life. This earlier explanation views the generative capacity to create variety in life as a special and valued property of life. It recognizes in the external environment both orderly change, to which biological life is adapted, and unpredictable and catastrophic change, which is destructive of life.

Thinking from the concepts of speciation and natural selection to those of biodiversity produces a potential ethical dilemma. If living organisms have the capacity to respond to environmental change, even of extreme form, by adaptation and evolution, why should we attempt to preserve the diversity of the Earth in its current form? Although some of the organisms may go extinct under human pressure, won't a corresponding number adapt to the new conditions humans create? We cannot resort to factual knowledge to answer this question, because we do not know the rates of adaptation and evolution for many species, nor do we understand adequately the environmental factors that cause extinction.

In the absence of such knowledge, we must base our action on ethical principle. Defense of biodiversity rests on the principle that all living organisms have intrinsic value. They also have instrumental value to the other members of the ecological communities in which they live, as well as to human economies. There is no justification for humans causing the extinction of other forms of life, either directly by killing them or indirectly through habitat change. We do not know enough about the biology or ecology of living systems to risk losing one. But the issue is larger than that. Speciation and natural selection are mechanisms that evolved to permit life to continue in an unpredictable future. To cause extinction and reduce this capacity is foolishness. We humans are not exempt from these processes. Human adaptation, which is cultural and technical as well as biological, should fit the environment as closely and as positively as that of other forms of life. Otherwise, our inability to adapt to and evolve with our environment will be accompanied by a loss of human well-being and health.

Readings

Ender, John A. 1986. *Natural Selection in the Wild.* Princeton, N.J., Princeton University Press.

Hutchinson, G. E. 1959. "Homage to Santa Rosalia, or Why There Are So Many Kinds of Animals." *American Naturalist* 93: 145–59.

Solbrig, Otto T. 1991. "The Origin of Biodiversity." *Environment* 33,no. 5: 17–20, 34–38.

Wilson, E. O. 1992. *The Diversity of Life.* Cambridge, Mass., Harvard University Press.

Cluster 4

Interaction Between Individuals and Species

In Cluster 4, the second cluster of concepts taking a biological perspective of ecological interactions and processes, we will once more shift our focus. Up to this point we have been analytical, taking a deeper and more intense view of the systems of interest. This reductive process has led us across two hierarchies, the first dealing with land-water systems and the second with population-individual systems. Now we will shift to a synthetic perspective, seeking to understand how to build up ecological systems from their parts.

As I discussed in the introductory chapters, the constructive point of view has specific attributes. It requires that we have a design which indicates how we will attach one part to the next and how these parts ultimately will contribute to the building of a whole. Further, it requires that we understand the parts sufficiently well to build with them. In construction terms, we need a parts list and a plan. In the language of technical ecology, we need assembly rules to build systems.

This perspective is culturally congenial to us. As builders of environments, we are familiar with the construction of systems. Indeed, it is a "natural" way for us to think. I frequently find students puzzled by my suggestion that they think reductively; reduction is contrary to the synthetic perspective they have lived with all their lives. Therefore, it is not at all strange that ecologists have taken a viewpoint that appears to be contrary to the traditional analytical approach of science.

Let us address this apparent contradiction from two points of view. First, although science describes the natural world and seeks mechanistic explanations for the patterns it discovers, it also builds theory. Theories are statements about regularity, which permit us to predict performance of the system under given conditions. Ideally, our theory must be grounded in testable hypotheses. If the theory withstands our tests, then we can reason from it and use it to construct systems. In this way we can be constructive, without assuming design or purpose. Second, ecologists take a synthetic view when they apply their experience and explanations by analogy to other systems of interest. This contextual perspective is commonplace to ecological science. Ecologists regularly move up and down hierarchies of systems. Probably, ecologists are the least constrained scientists of all because their systems of interest range across such a wide range of scale. Thus, the assemblage of ecological systems from subsystems becomes a test of our knowledge and understanding. If we can construct viable systems, we know that we understand some of their key structures and functions. And if we understand how to construct natural ecosystems, then we can apply this knowledge to the built environment and, one hopes, create sustainable systems there.

Cluster 4 will focus on the construction of systems of interactions built by placing two or more entities together in space and time. We begin with individuals and construct a population. The model we will use contrasts the positive and negative features of interactions. Both will be important. Then we will use species to construct biotic communities. These two levels, population and community, recreate the systems needed for the construction of ecological systems of land and water. In this way the circle of analysis and synthesis is completed and we see that ecology too fits the circles of relationships that we have projected on the natural patterns of the planet.

19

Interactions Between Individuals

In our exploration of ecological concepts using a hierarchy of organization, we reached a limit when we considered the individual organism as an ecological entity. Now we will begin the construction of ecological objects from individual organisms. The question addressed in this concept cluster is: What are the rules for assembling a collection of individuals into a population or social group? This is a difficult question and it will take several sections to suggest an answer.

The Interaction Matrix

To interact means for the lives of two entities to intersect in some way. At the extremes the intersection may be positive and benefit both, or it may be negative and injure both. A range of possibilities lie between these two extremes. A simple matrix has been used to illustrate these possibilities (fig. 19.1). Each matrix cell is characterized by a defining name or term.

In the interaction matrix two individuals face each other. There are three possible results of an interaction: positive, neutral, or negative. Thus, the matrix has nine possible interactions. If both individuals have positive interactions, the interaction is termed mutualism. Mutualism is a plus-plus case: both individuals benefit from the interaction. At the other extreme, if both lose, the interaction is called competition. This terminology seems contradictory to an American, because we are taught that positive ends come from competition. However, when we examine competition we will see that both individuals lose, although one may lose a great deal more than the other.

The neutral cell in the middle of the matrix means that neither individual gains or loses from the interaction.

Individual Two

		Positive	Neutral	Negative
Individual One	Positive	Mutualism	Commensalism	Predation Parasitism
	Neutral	Commensalism		
	Negative	Predation Parasitism		Competition

Fig. 19.1   An interaction matrix for two individuals or two species.

The case where one individual gains and the other loses is represented by dominance of one individual over another by predation or aggressive behavior. An example is the pecking order in birds. The dominant individual or individuals obtain access to food and shelter and prevent the subordinate individuals from receiving an equal share of resources. Their survival may be threatened as a consequence.

The final case is where one interacting individual gains or loses, but there is neither a positive nor a negative impact on the other member. This situation is called commensalism when two species are involved. Of course, it is difficult to determine the neutral condition. Can we ever be certain that an individual within a network of relationships does not benefit or suffer? I think not.

The interaction matrix gives us a language with which to talk about interaction. The following chapters will explore these patterns.

The Concept of Interaction

Interaction operates on two principles, proximity and relationship. Interaction occurs most often with individuals or species close at hand. The chance of being near another is more likely to cause an interaction to occur than are specific properties of the organism. The degree to which an interaction is positive or negative is a function of the relatedness of the individuals. Organisms are more likely to act positively to kin than they are to strangers. Because kin are likely to share genetic characteristics with the interacting individual, by acting positively to its kin the individual enhances the likelihood of its genes (those it shares with its kin members) contributing to the next generation.

Controlling Interactions

Organisms differ greatly in their capacity to control interactions. Many organisms are moved about by the force of wind or water and chance plays the largest role in governing where the individual will be located in space-time. Other organisms have the capacity partly to control the interactions. This capacity is best expressed as a continuum, with one end representing organisms that control access to resources in space-time. In this case the individual limits the likelihood of having to share resources with any other individual except its kin. The other end of the continuum represents elaborate social groups, where individuals have assumed special functions for the good of the whole group. In this case autonomy passes to the social group.

The control of interaction depends on the behavioral and physiological capacity of the organisms. Much interaction is concerned with the allocation of resources. For example, space is a resource and individuals partition it so that they can be assured of mates, resources, refuge sites, and so on. The area immediately around the organism is called the home range, if we are speaking of higher vertebrates. The home range is the area the individual knows thoroughly and travels over daily. It contains the food and shelter needed over the period of activity. But the home range is not fixed; it may expand, contract, or be moved.

A landscape patch may be divided into a number of home ranges. The ranges reflect the quality of the habitat. The dominant organisms claim the best home ranges and the youngest and weakest must make do with the poorer ranges. If the range is defended against neighboring organisms, it is called a territory. Territories are usually associated with nesting and reproduction. Subordinate or young individuals attempt to find a home range by traveling to new habitat or by establishing themselves in marginal habitat. The chance of survival in such situations is usually poor.

Other organisms do not exercise as much control over interactions as individual vertebrates. Rather, the survival of species is influenced more significantly by the number of young produced. If the parents produce many young, there is greater likelihood that some will find suitable habitat and can grow and reproduce. John Pinder and I and our associates have studied an example of this pattern in our investigation of the ecology of loblolly pine in grassy fields in South Carolina. The pine seed is produced intermittently. But periodically the trees produce large quantities of viable seeds that are wind-dispersed from the parent trees into the fields around the parent. When the seed germinates, it is subject to consumption by mice and insects and to competition with the other vegetation. Very dry summer weather conditions cause young seedlings to perish in the fields. But under the canopy of the parent trees offspring find suitable conditions for growth and survival. Older trees are often surrounded by circles of their offspring, which grow up through their parents' branches. We suspect that the moisture supply is slightly better in the canopy shade than in the open field and this condition permits seedlings to survive drought conditions. In this case the individuals act by chance. Those that fall near the parent canopy benefit in some way by the association, but the individual seeds have no choice where they fall.

Other organisms depend on physiological responses to resource gradients to control interactions. For example, Toshihide Hamazaki (1995) studied the greenhouse milliped (*Oxidus gracilis*), which lives in the soil-litter interface of forests and fields in temperate and tropical habitats worldwide. This species may become so abundant that it creates hazards on roadways. The milliped has a cuticle through which water passes readily, and therefore to survive the milliped must find moist conditions. Hamazaki placed wooden boards on the forest floor and found large numbers of millipedes sheltering under them during midday. Apparently these organisms move about following gradients of suitable food and moisture on the forest floor. Once a suitable site is located, the milliped stops moving. Where a particular milliped is at a given time is a matter of chance. Aggregations of millipedes are associated with suitable habitats. Thus, the apparent interaction of the millipedes is a function of their shared physiological requirements. Although the milliped can move, it does so in order to find suitable habitat, not to create and maintain social interactions between organisms.

Social Organization

Social behavior creates a problem for the ecologist. After all, according to the theory of evolution, organisms compete for scarce resources, for mates, and for space. If resources are limited, how could potentially competing individuals and species cooperate and through their cooperation build social relationships?

From a genetic perspective, social behavior among individuals must result in improved fitness (that is, in improved production of offspring). We can evaluate this possibility through an ecological cost-benefit analysis (Slobodchikoff 1988). The costs of social behavior are increased competition for resources, higher parasite loads, greater probability of misdirected parental care and infanticide, and increased conspicuousness to predators. These costs are balanced by benefits of greater defense against predators, more effective exploitation of resources, and a greater likelihood of finding a mate (Deag 1980). How the balance between these costs and benefits is decided depends on a variety of factors. For example, in deer, males may group together to enhance vigilance against predation, while females may group together to realize their investment in their daughters. Deer tend to form male groups and groups of adult females with their offspring. Daughters tend to stay with the adult female group, while young males leave the female group as they become sexually mature. During the breeding season males will defend groups of females organized into harems. In this example, genetic relatedness is not important in the social grouping of males, but it is required for the social grouping of females. One answer to the question asked above is that social behavior evolves to exploit resources or provide services that are unavailable to the individual. For example, lions tend to form groups and, by acting together socially, can stalk and kill prey more efficiently than they can as individuals.

If group behavior evolves, then we can ask, How large can a social group be? Groups vary greatly in size. For example, groups of sea birds on islands can number in the tens of thousands. The buffalo herds on western prairies were equally large and the passenger pigeon occurred in such large numbers that it was said to "darken the sky." C. N. Slobodchikoff and William C. Schulz (1988) suggest that aggressive behavior regulates the size of a group. At small size there can be high aggressivity, with males defending resources and mates. No individual can easily enter such groups. If the size becomes very large, however, aggression acts against survival and reproduction. In confined groups of mice, allowed to breed freely and provided food and water in excess, the populations may become so large that females trying to build nests have other females steal their nesting material when they are away from the nest feeding or finding more material. The consequence is that young are born on the ground without nests and their survival is very low. Further, under these conditions there is a great deal of random violence, with unprovoked attacks on subordinate males and females. At very large group size, the population cannot reproduce and it crashes to zero or has only a few survivors.

Another solution to the problem of controlling interactions is represented by the social insects. In these organisms individuals have given up their autonomy and the group functions as the individual. In the honey bee, for example, the queen is responsible for reproduction, male drones serve to fertilize the female, and the female workers feed and care for the eggs and queen and maintain the hive. There must be one queen in the hive, but the numbers of the other three classes of individuals or potential individuals can be quite different. A beekeeper can tell the health of his or her hives by the numbers of workers visible on the combs.

Sometimes a social group defends resources against other social groups. For example, the southern harvester ant (*Pogonomyrmex badius*) builds deep mounds in sandy soils. These mound systems may be several meters deep and extend down to the water table. The worker ants range over a wide area, traveling on well-defined paths. Ants from neighboring colonies are antagonistic and will attack one another. The consequence of this behavior is that colonies, marked visibly by large, vegetation-free mounds, are distributed in space on a reasonably well defined grid, which divides up space and the resources to satisfy the needs of each colony.

Clearly, organisms have used the interaction matrix in complex ways to satisfy their life-history requirements. Every imaginable arrangement seems to have been tried by some organism at some time.

Implications

In focusing on how individuals interact with each other, we enter directly into the realm of formal ethics. The ecological description of interaction provides patterns to organize our observations, but it does not tell us what we ought to do in a particular situation. This is the task of ethics.

Our ethics are taught to us by parents and social institutions. But by the time we reach school, we discover that we are expected to participate in multiple social interactions where we play different roles and that this activity is governed by mutually agreed upon social rules. The most practical rules are those used among kin; the most lofty are the social glue that keeps society functioning. But frequently we find confusing and contradictory expectations.

The human organism seems to be among the most social of animals. Few of us could survive alone for any length of time. Indeed, banishment or confinement to a solitary cell is among the most severe forms of punishment for breaking social rules. Yet we also have a propensity for dividing social groups and for finding means to compete with or to express a prejudice against the other. I suppose our emphasis on the individual and our carelessness about maintaining the social organism are due to the fact that we are social without choosing to be social. There is no way we can escape this property and it is so obvious that we can ignore it. It is always there. So, as much as we attempt to escape and express our individual selves, once we are successful at being independent we immediately form a new group and recreate sociality.

The proposition that we are social not by choice but by evolution means that social behavior is genetic. We assume that social behavior provides enhanced fitness in particular situations. Edward O. Wilson (1975, p. 3) even used sociobiology to explain the evolution of human ethics:

> The biologist, who is concerned with questions of physiology and evolutionary history, realizes that self-knowledge is constrained and shaped by the emotional control centers in the hypothalamus and limbic systems of the brain. These centers flood our consciousness with all the emotions—hate, love, guilt, fear, and other—that are consulted by ethical philosophers who wish to intuit the standards of good and evil. What, we are then compelled to ask, made the hypothalamus and limbic system? They evolved by natural selection. That simple biological statement must be pursued to explain ethics and ethical philosophers, if not epistemology and epistemologists, at all depths.

Michael Ruse (1985) evaluates this claim and tries to build a bridge between traditional ethics and sociobiology. He concludes that morality has selective advantage to the individual. The immoral individual fails to help relatives or friends and does not get help when help is needed. Although people are moral because they choose to do the right thing, not all human activity is voluntary—think about falling in love! Our legal system recognizes the involuntary with the defense of "an act of God." In this situation something in the *environment* exonerates the individual. Thus, although morality may have a recognizable selective advantage to an individual, it also can be shaped by environmental selection of genetically based traits. In this conclusion, Ruse brings us back to the familiar point where both genetics and ecology play a role.

The human paradox is that although humans can act individually, they always act within a social context. The paradox is at the root of all the great and small tragedies in the stories and songs of humankind. We can choose to act as if we could escape the social constraints of human society, but we cannot in realty do so. Ethics speaks to what we should do in this paradoxical situation.

Let us move from the social reality to the ecological reality of humankind. Just as we are unable to escape our social nature, so we are unable to escape our ecological nature. We are, like it or not, locked into a set of relationships with other organisms and habitats. For the human race as a whole, this means that we interact with the entire planet. There is no escape from this reality. Yet we can act as if we were free of ecology, as if ecological relationships did not exist. The consequence of that decision is harm to the individuals and societies which make that choice. Because we are locked into social and environmental relationships as a consequence of our being human and being a living organism on this planet, then our object should be to sustain these patterns of interaction as much as we can.

This objective is the foundation for an environmental ethics. An action is right when it builds, enhances, benefits, and maintains ecological and social interactions. An action is wrong when it breaks, interferes with, or destroys interactions and connections.

Readings

Kellert, Stephen R., and E. O. Wilson. 1993. *The Biophilia Hypothesis.* Covelo, Calif., Island Press.

Paine, Robert T. 1966. "Food Web Complexity and Species Diversity." *American Naturalist* 100: 65–75.

20

Mutualism

The cell of the interaction matrix (see fig. 19.1) labeled mutualism represents the case where both partners benefit from an interaction. We also may call mutualism cooperation and, when it involves the interaction of two species in a sustained biological relationship, symbiosis. All of these words indicate a positive result for both parties in the interaction.

A probably unfamiliar but extremely important example of mutualism is the relationship between plant roots and fungi that increase the efficiency of uptake of water and essential nutrients for plant growth. The fungi are called mycorrhizae, and they increase the surface area of the roots, help the roots contact a larger proportion of the soil system, and also help solubilize nutrients in soils where they would otherwise be unavailable. Mycorrhizal fungi are divided into two types. The ectomycorrhiza form a sheath of hyphae around roots and endomycorrhiza penetrate the cortex of the roots. Most plant families have members that exhibit mycorrhizal relationships; however, many early successional pioneer species are nonmycorrhizal.

The role of these fungi is best expressed in soils that are nutritionally poor. Under these conditions the symbiotic relationship is critical for plant growth and survival. The benefit of the fungi is greatest for nutrients that diffuse slowly through the soil, such as phosphate, ammonium, potassium, and nitrate. For example, transport of phosphate to the root through mycorrhizal fungi can be a thousand times faster than by diffusion through the soil (Chapin 1980). In addition, the fungi can directly alter the availability of nutrients at the root surface by various chemical processes, such as altering the root surface pH, solubilizing rock phosphate, and stimulating organic matter decomposition. The mycorrhizal fungi obtain organic products produced by the plant. Indeed, N. L. Stanton, in her discussion of grassland underground ecology (1988), says that endomycorrhizae are a "sink" for plant sugars. This source of energy provides the fungi an advantage that is not available to other free-living soil microorganisms.

The mycorrhizal fungal hyphae penetrate the soil at a distance well beyond that of an uninfected root and increase the surface area of roots many times. Fungal hyphae also form a link between living or dead organic material and live plant roots, thus providing a passage for nutrient transfer that is entirely under biological control. This link has been called "direct nutrient cycling." R. Herrera and his associates (1978) demonstrated direct cycling in the Amazonian tropical forest by showing that radioactive phosphorus moved from a dead leaf through the fungi to the roots in the surface litter. In addition to links with organic matter, fungal hyphae may connect with fungal hyphae growing on other trees, even those of other species, thus providing a potential link for flow of chemical information among individuals in a forest. This underground network is as important in tying the ecosystem components together as are the above-ground food webs discussed in Chapter 10.

Mutualism has been known to be significant in ecological relationships as long as the history of the subject, yet it has received much less attention than its opposite in the interaction matrix, competition. Paul Reddy's (1990) analysis of emphasis on these interaction concepts in ecology textbooks reported that the number of pages devoted to competition in twelve texts was 462 pages, compared to only 59 for mutualism. Yet, as Eugene Odum (1992) emphasizes, mutualism is the cement that holds ecosystems together.

There seems to be no logical reason why competition should be chosen over mutualism as a way of organizing relationships. Some deeper structuring phenomena appear to explain this difference in emphasis. I suspect that one of the factors is gender. Males in Western cultures often structure male society through competitive interaction. Because ecological science has been dominated by males, one might expect that male modes of interpreting nature would occur most often. One wonders how many mutualistic relationships might be discovered if we looked for them.

Of course, there may be many reasons for emphasizing one concept rather than another. Competition lends itself to the reductionist approach that is common in science. It seems easier to tease apart interactions into competitive pairs than to synthesize cooperation among pairs. Possibly of even more significance is the context in which studies of interaction take place. This context is the theory of evolution. For example, organisms might have an advantage by expending energy in cooperative activity if the recipient of the benefit shared genes with them. Cooperation would then increase the likelihood that their genes would influence the next generation. There would be less advantage to being cooperative with strangers. But if resources are limiting to organisms, how could individuals without an ability to predict the future choose a form of interaction that appears to handicap them in competition for resources in the immediate present? How could mutualism evolve?

Evolution of Cooperation

The conventional model of Darwinian evolution states (1) that individuals produce many more reproductive units or offspring than can survive, and (2) that these offspring compete with one another for resources so that only the fit pass through the filter of natural selection and produce the next generation. In the Darwinian model, competition is the mechanism used to guide or direct change. In the logical structure of the Darwinian model, cooperation might evolve by enhancing the fitness of the cooperators over that of competing individuals. Otherwise, cooperation would be selected against. Can we visualize a mechanism that could result in the evolution of cooperation?

R. Axelrod and W. D. Hamilton (1981) examined this problem and developed a hypothetical mechanism that might explain how two individuals or two interacting species could evolve a cooperative strategy. Their solution is not of general use, however. For example, it does not explain how fungi and roots evolved a symbiotic relationship, but it does propose an answer to our question for certain interactions.

TIT FOR TAT

Axelrod and Hamilton used the game of the Prisoner's Dilemma as the basis for their development of a theoretical answer to the question of the evolution of cooperation. The game is played as follows: Two prisoners (players) decide to escape. Each has two options: to cooperate and improve their chance of escape, or to betray the other prisoner and therefore gain a personal advantage in the prison. The payoff is in the form of accumulation of a score for the game. Betrayal yields a higher payoff than cooperation, but if both individuals betray the other, they do worse than if both cooperate. Here is the dilemma: individual betrayal yields the most reward but double betrayal yields less than does cooperation. Should the first player cooperate with or betray his fellow prisoner?

The game begins (fig. 20.1). Its rules are defined as $T > R > P > S$ and $R > (S + T)/2$, where the letters indicate cells of the interaction matrix, with the neutral category removed. If player A cooperates, the option for B is to cooperate, yielding R, or to betray, yielding T. Because $T > R$, it pays to be-

　　　　　　　　INDIVIDUALS AND SPECIES

A

Cooperation　　　　　Defection

	Cooperation	Defection
Cooperation	Reward for cooperation R = 3	Suckers payoff S = 0
Defection	Temptation to defect T = 5	Punishment for mutual defection P = 1

T > R > P > S　　　　R > (S + T) / 2

Fig. 20.1　The matrix for the Prisoner's Dilemma game. Two prisoners (A and B) face each other. Each has the choice of cooperating with his or her cell mate or defecting to the jailer. The payoffs for each alternative differ, as shown in the figure. The relationships between payoffs are shown below the matrix.

tray the other prisoner. If the player A chooses betrayal, the choice of B is to cooperate, which yields S, or also to betray, yielding P. Because P > S, it pays to betray. No matter what the other player does, choosing betrayal always seems to pay. How can cooperation pay?

To answer this question Axelrod and Hamilton ran an experiment. They asked game theorists to suggest strategies that would lead to cooperation in the Prisoner's Dilemma game and then played games using these strategies, making two hundred moves per test. They found one strategy, TIT FOR TAT, that led to cooperation more often than betrayal.

The TIT FOR TAT strategy always begins with cooperation on the first move and then repeats the move of the other player on all following moves. Thus, TIT FOR TAT is a cooperative strategy based on reciprocity. Axelrod's test showed that the strategy was very successful and robust, and resisted invasion of other strategies. Its success was due to the fact that although the player was never the first to betray, the player always retaliated after betrayal by the other but was forgiving if the other player became cooperative.

Axelrod and Hamilton suggest that TIT FOR TAT strategies could evolve into cooperation, especially if they were influenced by a bias toward relationship, such as kinship. Kin are more likely to respond to cooperation by further cooperation than are nonkin. But to use kinship it is necessary for the organisms to recognize degrees of relatedness. Male promiscuity, for example, reduces the likelihood of kin recognition.

The TIT FOR TAT strategy depends on the players meeting again. It also assumes that an individual cannot escape the game after competition without facing retaliation, and that the organisms can remember their partners and the partners' last move. If the cooperators combine and refuse to associate with the defectors, then there is opportunity for the cooperative strategy to spread.

Is there experimental evidence for TIT FOR TAT strategies? L. A. Dugatkin and M. Alfieri (1991) have shown that fish are more likely to associate with cooperators than noncooperators in experimental chambers. They provided small guppies with a potential confrontation with a predatory pumpkinseed sunfish. Certain fish, the inspectors who are surrogate cooperators, investigated the danger of the predator for the fish school. Inspection is risky because the investigating fish might be attacked and possibly killed by the predator. Yet inspection provides benefits not otherwise available to the entire school. If two fish approach the predator, both benefit (B) and both share the cost (C). The payoff is $B - C/2$. If one approaches but the other does not, the inspector bears all the cost but both benefit. If neither approaches, then the benefit is zero. In this experimental situation, if $C > B$ but $B > C/2$, then the test conforms to the TIT FOR TAT strategy.

The experimental chamber was constructed of plastic so that fish in three lanes could see one another but were otherwise separated. The predator was placed in a tank at the end of the three lanes. The center fish was able to select one or another of its associates on the basis of that fishes behavior to the predator. Dugatkin and Alfieri showed that guppies selected cooperators more often and that the fish remembered cooperator behavior for at least four hours. Here was experimental evidence that "everyone loves a cooperator" and moves away from a defector.

In this discussion of mutualism, I have mixed comments and examples from individual organisms and populations. Clearly, the interaction matrix and the tit for tat game is most clearly applied to the interaction of individual organisms that face one another within an environment. Axelrod and Hamilton's suggestion that closely related individuals are more likely to cooperate applies to the individual level of organization. In the mycorrhizae-root example, the relationship is between two species populations. This is a more complex system, involving the interactions of higher and lower plant populations and the interactions of individual organisms within these populations. The system becomes even more complex when we consider more than two species that might occupy different trophic levels. This level of complexity becomes difficult to understand, because a competitive interaction between two members of a community might have an indirect positive impact on other interlinked pairs in the community. As Paul Reddy (1990) implies, we are involved in mutualistic interactions everywhere we look.

Although mutualism is of comparatively little interest to modern American ecology, it has been the focus of attention in other countries. For example, the Russian prince Peter Kropotkin, who wrote a book in 1902 titled *Mutual Aid: A Factor in Evolution,* opposed the competitive emphasis of Charles Darwin in his *Origin of the Species.* E. P. Odum (1992) comments that Kropotkin's most important contributions included identifying two forms of the struggle for existence. One involved organism against organism, which leads to competition and the other involved organism against environment, which leads to cooperation. Odum emphasizes that organisms don't struggle against or fight their physical environment. They adapt to it, and adaptation might include development of cooperative behaviors.

Implication

The discussion of cooperation in terms of the TIT FOR TAT solution to the game of the Prisoner's Dilemma may be satisfying to the scientist looking for a mechanistic explanation for the evolution of cooperation, but it is unsatisfactory to someone seeking to understand mutualistic behavior. Neither the discussion of competition nor cooperation in mechanical language has the emotion, strong feeling, and potential for violence or affection that are the central facts of social interactions. Ecological explanations fall far short of helping us understand these common human traits. However, they are of value in explaining social interaction in terms of other ecological phenomena and organisms, fitting mutualism into (more or less) a universal ecological explanation of life.

In my view, the need to create an explanation of mutualism to fit the broader pattern of ecological interaction has turned the facts upside down. Although Darwinian evolution may be useful in explaining development in individuals, it lacks the power to explain group behaviors. Individuals may compete for resources, but the same individuals may cooperate to improve their access to resources at another time or place. Individuals are not abstract automatons; they move back and forth across the interaction matrix, guided by current evidence and past experience. Odum's suggestion that organisms cooperate when facing an external challenge but compete when concerned with resource allocation helps open the options for a balanced view of these opposing interactions.

Individuals are more likely to cooperate when they are familiar with other members of the group. Family or kin groups are loci of cooperative behavior that become a model for widespread behaviors. The social basis of interaction in humans is only being slowly understood through the work of the social sciences, and this work has not been applied to ecology. Our understanding of the interaction of competition and cooperation in organisms awaits development of an adequate understanding of relations in this larger context.

Reading

Axelrod, R., and W. D. Hamilton. 1981. "The Evolution of Cooperation." *Science* 211: 1390–96.

Boucher, D. H. 1985. *The Biology of Mutualism*. New York, Oxford University Press.

21

Competition

I have characterized biotic interactions in a simple matrix where two members of a pair are contrasted with each other and the outcome of the interaction is judged to be positive, neutral, or negative in terms of a particular life history or resource allocation strategy (see fig. 19.1). In this chapter I will focus on competition.

The interaction matrix provides a model for examining competition in a general sense. Competition may occur within any group of biological entities, individuals, populations, species, or communities. Obviously, competition between individual organisms is different from competition between populations because individuals and populations have different capacities and properties. In the individual, competition may cause slower growth, delayed reproduction, and/or death. In the population, competition may change density, productivity, and the capacity to resist change or stress. Competition has indirect effects, as well. For example, the direct impact of competition on individuals may have an indirect impact on the population through change in individual behavior. These complications make discussion of interactions confusing, because we may change the biological subjects as we develop the topic.

As noted in the previous chapter, of all the potential interactions between organisms, competitive relationships have received the most attention. Probably the major technical reason for this bias is the biologist's special interest in Darwinian evolution. The evolutionary paradigm assumes that natural selection operates through competing genomes. Ultimately, the individuals—which are the phenotypic expression of the genome—that leave the largest number of surviving and reproducing offspring have the strongest influence on the future of the population.

In this genetic sense, as in an economic sense, it appears that competi-

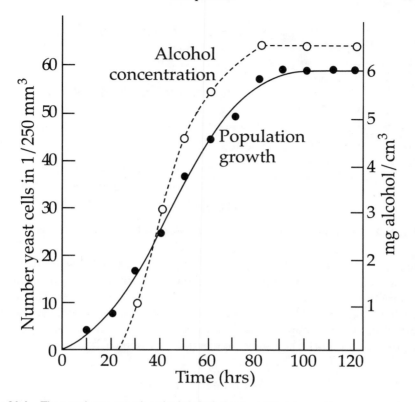

Fig. 21.1   The population growth and ethyl alcohol accumulation in yeast.

tion is a positive process in population survival. Why then is competition defined as the cell in the interaction matrix where both individuals suffer some loss?

In 1934, the Russian ecologist G. F. Gause developed a rule that two species cannot share the same niche—in other words, that no two species can do exactly the same thing. If there is not some difference in the way the species interact, they will compete and one will dominate and replace the other. This principle is called the competitive exclusion concept.

Gause (1932) based his concept on studies of populations of yeast cells grown in the laboratory (figs. 21.1 and 21.2). The yeast exhibit the familiar logistic growth pattern of biological populations (fig. 21.1). As they grow the yeast produce ethyl alcohol, and as the alcohol increases it becomes limiting and results in the population-growth curve reaching an asymptote. When grown alone, the two yeasts behave differently, have different values of r and K, and reach different population equilibria (fig. 21.2). When the

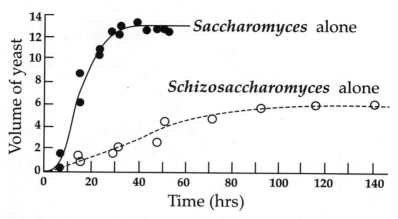

Fig. 21.2   Population growth of two cultures of pure yeast in Gause's experiment.

two yeasts are grown together (fig. 21.3), the growth curves are different. Both populations reach lower population equilibrium sizes when they are competing in the same media (compare figs. 21.2 and 21.3). This is the basis for saying that the interaction matrix yields a negative-negative relationship. Neither population does as well (reaches as high a population equilibrium point) when they are together as when each is alone.

Environmental Impacts

The competitive relationship also is sensitive to the environment in which it occurs. An elegant study of the effect of environmental factors on competition was made using flour beetles by the University of Chicago ecologist Thomas Park. Flour beetles make their burrows in flour and also eat flour. This insect is a severe pest in flour mills and warehouses, but these characteristics also make them relatively easy to culture and study. Experimental chambers containing flour can be emptied of flour, the flour can be sifted, and the entire population of beetles can be recovered, counted, and then introduced into a new chamber. Park and his associates made major contributions to our understanding of population ecology through their studies of these obscure insects.

Flour beetle populations are sensitive to temperature. Park (1948 and 1954) showed that a shift in the temperature of the flour medium directly influenced the competitive relationships between two species of beetles. The populations at 29.1 degrees centigrade fluctuated in numbers but maintained a dominance of *Calandra oryzae* over *Rhizopertha dominica* for 180 weeks (fig. 21.4). A change of only 3 degrees centigrade shifted the species dominance, with a great increase in *Rhizopertha* and a decline in

Saccharomyces

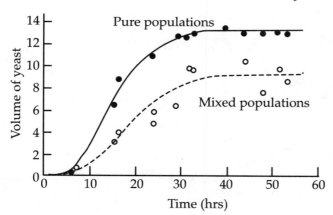

Shizosaccharomyces

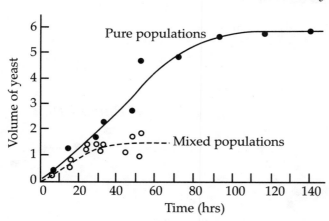

Fig. 21.3 Growth of populations of two yeast, *Saccharomyces* (top) and *Shizosaccharomyces* (bottom), in pure and mixed cultures, in Gause's experiment. The data illustrate depressed growth in mixed cultures.

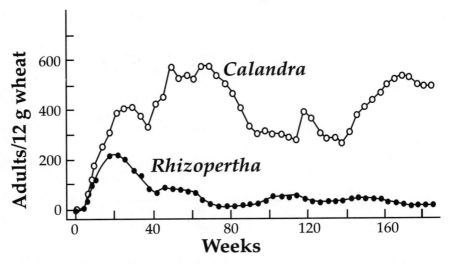

Fig. 21.4 Population trends in two flour beetles living together in wheat at 14 percent humidity at 29°C. After Birch 1953.

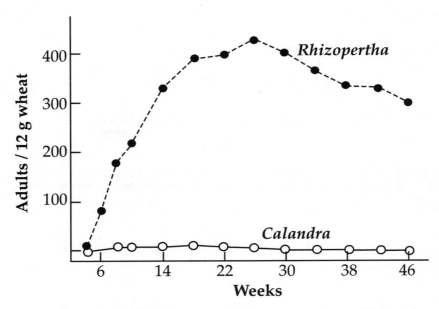

Fig. 21.5 Population trend of two flour beetles living together in wheat at 14 percent humidity and 32°C. After Birch 1953.

Calandra (fig. 21.5). The competitive relationship was completely reversed under these environmental conditions.

The Natural Environment

These laboratory populations demonstrate the competitive relationship clearly, but the organisms are living in simple experimental environments. Do the patterns appear in complex natural environments as well? The English ornithologist David Lack compared the niches of closely related pairs of British passerine bird species, and his data help us to address the question of competitive exclusion in the field. Lack arranged the niches of British passerines in the following pattern and placed the species into each category:

Niche Separation Factors	Number of Pairs
Geographic separation	3 pairs
Habitat separation	18 pairs
Feeding habits separation	4 pairs
Size differences	5 pairs
Different winter ranges	2 pairs
Apparent niche overlap	5–7 pairs

Lack was able to separate thirty-nine pairs of birds by five criteria. Five to seven species could not be assigned to a single niche separation factor. Most pairs of birds were separated by habitat; the other factors had approximately the same number of pairs. Lack's study shows that species separate themselves and reduce competitive exclusion in many ways, but use of different habitats is probably the most important.

Lack's broad-brush approach to competitive exclusion in closely related British passerine birds can be contrasted with a microstudy of related birds in a single forest stand. Robert MacArthur (1958) examined the niche separation of warblers in boreal forest. Five species of small, insect-feeding *Dendroica* occur in these forests and feed on the same trees. MacArthur asked how five species with similar habits could coexist. None of Lack's separation factors appears to be applicable to this situation.

MacArthur subdivided the trees into height zones and then identified a horizontal zonation on the branches of the trees, consisting of the base of a branch, the middle, or the tip. Thus, MacArthur recognized a subdivision of the habitat into vertical and horizontal microzones. By careful observation he found that the species feed in these different zones of the tree canopy (fig. 21.6). That is, the species had separate ecological niches at a microscale. Niche separation allowed the five species to live in the same

Height zones (10-foot units)

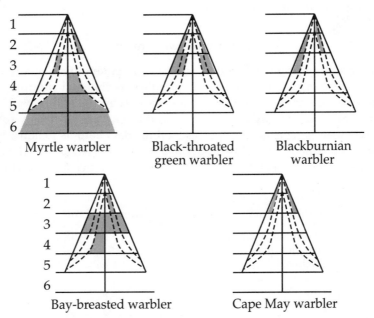

Fig. 21.6 Feeding positions of five species of warblers in coniferous forest in the northeastern United States. On the left of the "tree" in each graph is shown the number of seconds of observation of a species; the right shows the number of observations. The zones of the most concentrated feeding are shaded. Based on MacArthur 1958.

habitat even though they fed on similar organisms. This example illustrates the fine separation of the resource gradients possible in the natural environment.

An underlying assumption in these examples is that the species have sufficient time to interact with each other to reach a stabilized relationship. Competitive exclusion would be less likely to occur in unstable environments where the species never reach equilibrium. In these situations the populations compete for resources but cannot reach a balanced relationship, because the resources are changing rapidly in kind and amount. In these situations the species interactions are overwhelmed by the environmental relationships.

In this discussion the examples have been expressed as interactions between species populations and not between individuals within a species. Ecologists have often studied competition between species and examples are readily available for these cases, although behavior ecologists focus on competitive and cooperative relationships among individual organisms.

Ecologists have been especially interested in learning how different populations using joint resources survive in nature. The same patterns provide a basis for understanding the interactions of individuals, but we must remember that the endpoints are different. Individuals compete and the successful individual is able to grow faster, reproduce earlier, or control a larger territory than its competitor. The successful individual can garner more resources and leave a larger number of offspring and thus transmit its traits to more individuals in the next generation.

Implications

Competition must be evaluated within a larger context of the social group, the population, and the ecological system. Although competition may have a negative impact on both competitors in the interaction matrix, competition may have positive effects on the larger system of which the competitors are a part. In other words, the judgment about positive and negative relationships given by the interaction matrix may be misleading if it is not interpreted heterarchically.

For example, competitive relationships probably underpin the development of species diversity. The pressure of competition leads to an efficient partitioning of the resource base by sorting out inadequate competitors from the species pool. In the same way that economic competition in a market is an efficient way to set prices and allocate benefits and costs among producers and consumers, competition provides an efficient selection of the species potentially available to that habitat.

Thus, the interpretation of relationship depends on the frame of reference from which the interactions are judged. Sometimes, what are judged as positive or negative interactions at one level have an opposite impact at another level. It is commonplace in human affairs to see individuals who act to maximize their personal rewards in competition with others and by doing so injure the common good upon which the social contract allowing competition depends. Garrett Hardin (1968) called this "the tragedy of the commons." Individuals seeking to obtain the maximum personal benefit from a common resource ultimately destroy the resource. One solution to the tragedy of the commons problem is to devise commonly acceptable rules that limit individual behavior. What rules form in ecological systems to prevent injury to the system from competition?

I think that the answer to this question is that biological diversity and environmental variation mediate indirectly the injurious impacts of interactions in ecological systems. For example, the physical environment is changing in complex patterns over space and time. The broad planetary patterns that I discussed in the first chapters are chaotic on the biome to individual scales of spatial and biological organization. Organisms are seldom in fixed, stable relationships with the physical environment, although some species have evolved elaborate mechanisms to buffer them from these fluctuations. Environmental change not only limits the performance of organisms but also has a direct impact on competition and other interactions. Organisms respond to environmental change genetically, physiologically, and behaviorally. They have the capacity to relate to one another as well as to back away from each other. Selection may move organisms in one direction or another depending upon the circumstances. Thus, the interaction matrix is held in constant tension between a varying physical environment and the adaptive response and evolution of a diverse biota. Equilibrium interactions may form in boreal forests, the ocean depths, deep lakes (such as Lake Baikal), and the tropical rain forest; these places are precious and need protection. But most of the Earth surface is disturbed and dynamic, and in these habitats interactions probably act quite differently. We know little about these patterns, however, and they are especially difficult to study.

Humans do not usually interpret their interaction with the natural world in terms of competition between species, although that is what it is. We seek to remove competitors for crops, livestock, forest trees, and wildlife by killing the organisms that feed on these domesticated or human-valued species. The human goal seems to be a sanitized world in which no competitors to human desire exist. The program of conversion of a biologically diverse world into a simple garden serving human needs and desires has failed for at least three reasons. First, attempts to kill individuals or populations have seldom succeeded completely, with the consequence that some individuals escape extinction. These surviving individuals have been driven through a process of natural selection to resist the poisons invented by humans. Although the populations of these human competitors have been negatively affected for a time, most species that have been attacked have been able to recover. Organisms resistant to the poison or control procedure develop and reestablish the competitive relationships.

Second, humans are biological organisms and are also susceptible to the poisons or biocides used against other living organisms. The environment is interconnected, so that poisons placed in the environment to control competitors eventually also cause human ill-health and death. Because the biosphere is connected, everything that is thrown away comes back some time, somehow, like a boomerang. This means that the foods and products using chemicals that are dangerous to the so-called pest species are also dangerous to humans. The costs of protecting the public from cheaters and of discovering new ways for agribusiness to continue to use poisons are enormous. The purpose of control is defeated by the means of control.

Third, destruction of one species in ecological systems because it is a human competitor is disruptive to the system as a whole. Breaking the linkages of food webs and cycles of relationships disturbs the ecological system, and it is not easy to predict the consequences. Frequently, actions taken against competitors create other problems in the environment that produce larger and unexpected costs than the competition did in the first place.

The problem is that greed and incompetent management cause human demand constantly to push against supply. Food and material crisis is continuous. Why are humans so different from other forms of life, which fit into complex patterns with seemingly little effort? If humans could manage themselves so that stores of food and resources were available for times of need, they could fit into the Earth rather than being against the Earth. A rational adaptation of humans with natural organisms could evolve. Competition could then find some middle ground, where natural organisms and humans shared production and natural processes could be used by humans to reduce the effects of pest populations. Under these conditions the extreme forms of pest and disease outbreaks probably would be less likely.

This change, however, would require that humans adapt to the land and reject the paradigm of separation from nature. The "green" hope is that we give up the extreme forms of human competition, especially warfare, which are symptoms of deeper problems, and build cooperative societies, each adapted to its habitat and operating within its own interpretation of the human-nature relationship. It is my conviction that hope for a better human existence lies in this direction.

Readings

Harden, Garrett. 1968. "The Tragedy of the Commons." *Science* 162: 1243–48.

Law, R., and A. R. Watkinson. 1989. "Competition," pp. 243–84. In J. M. Cherrett, ed., *Ecological Concepts.* Oxford, Blackwell Scientific Publishers.

MacArthur, R. H. 1958. "Population Ecology of Some Warblers of Northeasterm Coniferous Forests." *Ecology* 39: 599–619.

22

Predation

Predation is a process in which one species feeds on another. In the dictionary predation is defined negatively, as plundering or preying on others. The word *predation* is derived from the Latin word for prey. Ecologists, however, have tended to use it in a more general way to refer to the impact of one species on another through feeding. Thus, predation may involve one animal eating another animal in the traditional sense, or it may involve the consumption of green plants, and be called herbivory. Organisms that eat both plants and animals practice omnivory.

Predation is of interest to ecologists because it may influence the distribution and abundance of the prey. It also may be a selective force leading to cryptic coloration or other adaptations, and it may structure communities. We will consider its role in organizing communities in a later chapter.

Predation and Populations

A classic story of predators and prey was developed by the distinguished British animal ecologist Charles Elton from the records of the Hudson Bay Company. The company collected fur from a variety of species across northern Canada for many years. From these data Elton and M. Nicholson (1942) described the fluctuations of snowshoe hare populations over fifty years (fig. 22.1). This species showed distinct patterns of increase and decrease over a broad region. On a frequency of about ten years, a period of increase was followed by a period of decline in population numbers. This is a cyclic pattern of population numbers, and although it is characteristic of several northern species, it is not common.

The snowshoe hare is prey for the Canada lynx, and Elton and Nicholson also graphed the catch of lynx pelts in the same time interval. There are

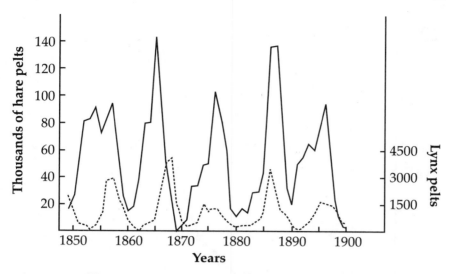

Fig. 22.1   Thousands of snowshoe hare and Canada lynx pelts taken at Hudson Bay Company sites from 1850 to 1900. Data from MacLulich 1957 and Elton and Nicholson 1942.

fewer lynx pelts, but they follow the same cycle as the hares. The two populations appear to be coupled by their relationship to each other or to their joint response to some external factor.

As we might expect, there has been a lot of speculation about these data. Some have suggested that the pattern is through the hare-vegetation food link. As the hare populations expand, they consume the vegetation. Eventually the food supply becomes inadequate and the population crashes, allowing the vegetation to regrow and provide the conditions for another upswing in the hare population curve. In this scenario the lynx population lives off the hare population and consequently the populations cycle together.

Lloyd Keith (1963) made a careful study of these data and concluded that if one does not demand strict regularity, the hare and lynx populations represent cycles of abundance and scarcity. We could interpret this pattern as a coupled oscillation in population numbers of predators and prey. It is not too difficult to imagine that a predator which fed almost exclusively on one prey species would fluctuate with its food supply. But is the predator controlling the prey or is it responding to the abundance of its food? If a predator had alternative foods, could it control the numbers of prey by concentrating its feeding on a vulnerable species?

These are important questions and the predator-prey relationship has been the subject of many studies and much speculation. For example, deer

INDIVIDUALS AND SPECIES

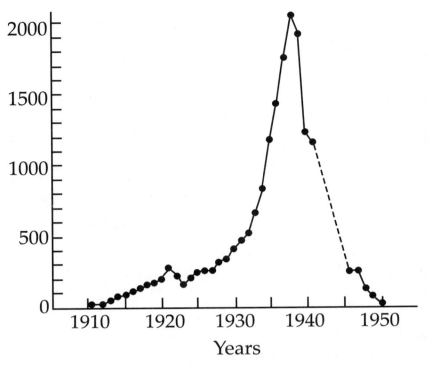

Fig. 22.2 Population growth of reindeer on Pribilof Island, Alaska. After Scheffer 1951.

populations have increased with the availability of habitat in many places. This species has the potential to increase to very high numbers, following the pattern of the J-shaped curve of population growth (fig. 22.2). At high numbers the deer are capable of destroying their habitat. Usually, if deer are unmanaged and there is no predation, the populations grow so large that they become vulnerable to a controlling factor such as an unusually cold winter or a disease. These mortality factors lead to widespread die-off, representing the crash phase of the growth curve.

It is common for people to attribute this growth pattern to lack of predators. They propose that if the wolves and mountain lions that fed on deer had not been killed, the deer herds would not have grown out of control. This is the position of the famed Aldo Leopold in his book on game management. Is this supposition true or not?

In the case of wolves feeding on caribou in the Arctic, A. T. Bergerud (1980) found that predators alone held the populations at about 0.4 animals per square kilometer, but that herds without predators increased to 20 or more per square kilometer. They would then suffer starvation losses. With predation the intrinsic rate of increase (r) was 0.02; with no predators it increased to 0.28. Bergerud's observations support the argument that predators can control prey populations.

Although this example suggests that predators can help maintain the health and condition of a prey population, predators also can be extremely hurtful to a population. The sea lamprey, an exotic species, was accidentally introduced into the Great Lakes, where it eliminated the lake trout fishery. This predator moves into streams to spawn and then returns to the lake, where the adults attach themselves to the sides of fish, eat a hole in the side, and suck out the body fluids. The lake trout population declined to essentially zero within twenty years after the sea lamprey was first observed in the lakes. The lake trout was not adapted to lamprey predation and succumbed to this new organism in its environment.

Prudent Predators

Where populations of predators and prey have lived together for a long time, they often have developed adaptations that permit both to coexist. Where predators have an extreme impact on their prey, it often occurs in a new relationship or in a highly stressed and disturbed environment where the evolved patterns cannot operate. The wolf is an example of a prudent predator, the sea lamprey of an imprudent predator. A prudent predator does not consume the prey in the age groups or the times when reproduction occurs. It feeds only on animals that would die under other circumstances and that contribute little to the productivity of the prey population. Usually, these prey are the old, the young, or the sick and injured. Predators associated with aggregations of prey move among the prey without attacking vigorous animals. But unusual behavior or an unusual gait, which would indicate weakness, immediately attracts the predator's attention. For example, on Isle Royal in Lake Superior, R. O. Peterson (1977) found that wolf predation on moose varied with age of prey. Wolf kill of older animals was pronounced (fig. 22.3).

Group living serves as a defense against predation in many species. Not only are there more eyes watching predators, but the milling crowd is confusing to the predator, which must focus on an individual to capture it. Then too, several prey animals frequently will fight off a predator.

Active predators, such as lions and hawks, are not successful each time they hunt for a prey animal. Actually, the frequency of success may be rather low—only about 10 percent. Because predation is not very efficient, predators choose prey that is weak or sick and concentrate hunting in locations where they have been previously successful. A redtail hawk, for example, will frequently return to the location where it was last successful and start to hunt from there. If a prey population is abundant, it will con-

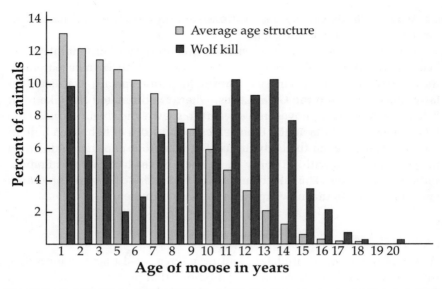

Fig. 22.3　Age distribution of 307 adult moose killed by timber wolves on Isle Royale, Lake Superior, from 1958 to 1974 and the age structure of the Isle Royale moose herd. Data from Peterson 1977.

tinue to attract the predator until the population has been reduced to a size where the frequency of capture is low. Then the predator searches for a new center of prey abundance.

Predators, disease organisms, and parasites are organisms that are coupled to other organisms which serve as their food and sometimes their habitat as well. As a consequence, the predators' life history is closely coupled to that of their prey, even to showing cycles of abundance linked to the prey cycles. The tightness of this coupling varies widely among organisms. Yet the predator or disease organism is faced with the dilemma of success. A high success rate, or an efficient adaptation to predation, can result in the death of the prey population and the demise of the predator. The predator must be prudent. How does prudency evolve?

Following concepts presented in Chapter 18, selection eliminates predatory organisms that destroy their food or habitat. Virulence destroys the host and the disease-causing organisms too. Prudency requires a balanced relationship that may depend on the presence of refuge sites for prey, a coupling of life history events so that the prey produces offspring which overwhelm the predators' capacity to control them, slower rates of response of predators to environmental change, and so on. Long-term predator-prey relationships always have evolved some form of prudency. Where we find predators or disease organisms killing the host, we often are dealing with newly evolved relationships or to highly disturbed environments where the prudent patterns can no longer operate. A predator introduced to a new habitat is frequently as terrible as the sea lamprey was in the Great Lakes.

The principle that emerges from this discussion is that predators influence but do not control their prey in relatively undisturbed environments where a long-term relationship between the couple can evolve. However, in highly disturbed environments or when a predator is introduced into a new environment, predators have not evolved prudent relationships and they can cause the prey to decline to very low levels or even go extinct. These latter events are not always associated with human disturbance. In paleohistory the appearance of advanced predators caused the extinction of unadapted prey species.

Implication

In my experience, several phenomena cause members of Western cultures to act completely emotionally, without thought of consequences. One of these is the discovery of gold. A gold strike causes otherwise rational people to throw moral behavior, honesty, prudence, and peacefulness to the winds, to the great cost of indigenous people, families and friends, and the environment. A second is an encounter with a predator. The predator is destroyed immediately, without thought or reflection. Our culture teaches us that it is good to react in this way. The encounter almost always elicits a fear of being attacked, and if the predator is dangerous enough—a spider, a snake, a hawk, a grizzly bear—then we back off and wish that we had a weapon to carry out our deadly intent. A policy to eliminate all predators would probably meet with general favor.

Why do we have such irrational fears? One of the prime reasons must be ignorance. We know very little about these organisms, even from a scientific and natural history perspective. The average citizen who never encounters nature except on a visit to a park knows much less. Predators are not abundant. They occupy the top part of Elton's pyramid of numbers and they require searching out by the ecologist. Rather, I think that the problem is that we invest predators with human properties. We use them to carry fears and dangers of a human sort. This projection is deeply placed in the culture. Children are taught the stories of Little Red Riding Hood, the serpent in the Garden of Eden, and the lion, tigers, and bears of the Land of Oz. A hunt club evening always includes predator stories and our character is defined by such adventures. The predator occupies the position of the other, the outsider, the dangerous wild one. We can project our fears onto it, we can shiver with delight when we think about it, and we can destroy it and do good, defending civilized life, domestic livestock, and our home. It is a wonder that the predator can carry so much human baggage.

If we move from this atavistic viewpoint to one that is rational and based on our limited knowledge of ecology, then predation raises other questions that challenge our understanding and our ethical positions derived from ecological knowledge. For example, the wolf that takes the young, old, and diseased caribou kills individual animals. If we place value on the individual animal, as we do if the prey are sheep or cows, then we interpret predation as bad. However, the wolf also removes animals that do not contribute to the herd's productivity and health. Thus, predation is good from the perspective of the herd. Our attitude about predation in general or in a specific case depends on the way we view the value of an individual prey organism. By removing predation, it is likely that ultimately we will cause more misery in the prey as the population expands, destroys its habitat, and then dies in massive starvation events. Tender-hearted removal of predation, including human hunting, can lead to terrible consequences. Which is worse?

Humans who live close to nature have complex and even contradictory answers to this question. Their attitudes toward predation may involve both competition and outright fear. Humans also act as predators and resent having to share a resource with other species. A degree of loss to predators is expected, but the tolerance for losses above this level is low. Predators also may attack and kill humans. The African villager views the lion as a danger but also, in a positive sense, as a source of power (Ngwa 1992). To hunt the lion requires special permission and there are special preparations for the hunt. To kill a lion is a courageous act and earns the hunter rewards, which are symbols of his courage and position in his culture.

The predator plays an important symbolic role in African cultures, representing ambiguous attitudes toward violence and warfare. The presence of both the predator and the warrior can be positive good but also, under other circumstances, a great danger. In these cultures, human society is balanced between positive and negative forces, just as are all other organisms. The predator represents this balance and tension.

Readings

Brooks, John Langdon, and Stanley I. Dodson. 1965. "Predation, Body Size and Composition of Plankton." *Science* 150: 28–35.

Hollings, C. S. 1959. "The Components of Predation as Revealed by a Study of Small Mammal Predation of the European Pine Sawfly." *Canadian Entomologist* 91: 293–320.

Mech, David L. 1970. *The Wolf: The Ecology and Behavior of an Endangered Species.* Garden City, N.Y., Natural History Press.

23

Coevolution and Niche

Coevolution is an evolutionary change in the individual traits of one population in response to a second population, followed by a reciprocal evolutionary response by the second population to the first. The species coevolve because each depends on interaction with the other to enhance its own survival and reproduction. Coevolution probably requires a long-term relationship, with each species performing actions that benefit the other, so that eventually a mutual dependency occurs. Coevolution is, therefore, a special form of mutualism. The plus-plus relationship is sustained over a sufficient length of time that genetic change occurs in both members of the coevolutionary pair.

Actually, it is quite difficult to find examples of coevolution. The problem is that we require a genetic response in both members of the pair, and it is rare to find a case where such a response is clearly demonstrable. In order to show coevolution, it is necessary to have an interacting pair of species, short generation times, and genetic markers, so that a change in the genome can be followed by the geneticist. Yet even though it is difficult to demonstrate coevolution in a pair of species, the concept has general value. Coevolution has been used as a synonym for interaction, symbiosis, mutualism, and plant-animal interactions—all misuses of the concept.

The word *coevolution* was coined by Paul Ehrlich and Peter Raven, director of the Missouri Botanical Garden, in 1964 in a paper on butterflies and their foods. They found that the feeding relationships of butterfly larvae are often fixed and predictable. The larvae are attracted to and consume very specific kinds of plant foods. Plants have evolved specific chemicals, called secondary plant compounds, that are not essential to their metabolism and are not inimical to their growth but influence the palatability of their tissues. Plants with these compounds experience lower feeding pressure from herbivores. The quantity of secondary plant compounds varies among individual plants and according to time of year and state of nutrition. If a herbivore, such as a butterfly larva, consumes these plants when the chemical titer of secondary plant compounds in the tissues is lower than usual, the insects might have an advantage over other larva that ordinarily would be repelled. If the butterflies provide some service, such as pollination, that enhance the plant's competitive position, there is a basis for a coevolutionary relationship in which each profits by associating with the other.

A classic example of coevolution was published by Daniel Jansen (1966), a distinguished tropical ecologist from the University of Pennsylvania. For his doctoral thesis, Jansen studied the coevolution of ants and acacia trees in Central America. In dry habitats in this region, swollen-thorn acacias are frequently found in hedgerows around fields or along roadsides. The enlarged thorns typical of these plants are inhabited by ants. The trees also have enlarged floral nectaries that produce a sugarlike compound and modified leaf tips that are eaten by the ants (table 23.1). The trees also are active year-round. When the ants build up a colony of about 1,200 workers on a tree, they patrol the tree and aggressively attack any herbivore that lands on the leaves or branches, protecting the tree from herbivory. They also attack any living competing plant that touches the acacia's foliage or grows within a 10–150 centimeter zone around the trunk. The acacia is thus able to grow rapidly, free of competition and herbivory. Acacias unoccupied by ants show severe defoliation by insects and loss of growing shoot tips. This impact can cause death or at least stunting, which leads to competition with other low-growing plants. For success, the acacias require the ants.

The ants also require the acacias (table 23.2). The acacias provide food throughout the year and a constant supply of new nest sites in the thorns. The ants could not establish such large and rapidly growing colonies if they were forced to forage on the ground.

Jansen comments that this is a case of an evolutionary feedback system par excellence. Evolutionary changes have occurred in both the plant and insect, enhancing their individual success.

Coevolution produces sets of species that are mutually dependent and linked less closely to other sets of species within the community. These sets of species are often called guilds, indicating their special relationship to each other.

The Niche

The concept of the ecological niche is widely used to describe the place of a species in its environment. It is helpful to bring it into the discussion at this point, because niche involves both the organism's interaction with its

Table 23.1   A comparison of general features of acacias and coevolved traits of swollen-thorn acacias. From Jansen 1966.

General Features	Coevolved Traits
1. Woody shrub or tree	1. Woody, with high growth rate
2. Reproduce from suckers	2. Rapid, year-round sucker production
3. Moderate seedling and sucker mortality	3. Very high unoccupied seedling and sucker mortality
4. Lives in dry areas	4. Lives in moist areas
5. Sheds leaves in dry season	5. Year-round leaf production
6. Shade-intolerant, vine-covered	6. Shade-intolerant, vine-free
7. Bitter-tasting foliage	7. Bland-tasting foliage
8. Foliar nectaries	8. Enlarged foliar nectaries
9. Compound, unmodified leaves	9. Leaflets with tips modified into Beltian bodies
10. Seeds, dispersed by water, gravity, and rodents	10. Seeds, dispersed by birds
11. Not dependent on another species for survival	11. Dependent on another species.

Table 23.2   Comparison of general features of *Pseudomyrmex* ants and coevolved traits of obligate acacia ants. From Jansen 1966.

General Ant Features	Coevolved Traits
1. Fast, agile, not aggressive	1. Very fast, agile, aggressive
2. Good vision	2. Good vision
3. Independent foragers	3. Independent foragers
4. Barbed sting sheath not inserted	4. Barbed sting sheath often inserted
5. Prey retrieved entire	5. Prey retrieved entire
6. Ignore living vegetation	6. Maul vegetation touch acacias
7. Arboreal colony	7. Arboreal colony
8. Highly mobile colony	8. Highly mobile colony
9. One queen per colony	9. Sometimes more than one queen
10. Colonies small	10. Colonies large
11. Diurnal activity outside nest	11. 24-hour activity outside nest
12. Few workers on plant surface	12. Many workers on plant surface
13. Discontinuous food source	13. Continuous food source
14. Not dependent on another species	14. Dependent on another species

physical environment and its positive and negative interactions with other organisms. Niche is more than habitat, although it is sometimes used to refer to the place and environment where an organism lives.

The concept of niche was invented by several ecologists. Among these was Charles Elton (1927), who made the helpful suggestion that the niche was the "profession" or role of a species. In this sense, niche includes all the interactions in a species' life history. In this definition we return to Dwight Billings's diagram of the environment of a single plant (see fig. 1.1). If the biotic element in the center of the diagram is a species, the niche is defined by all the interactions described in the figure. G. Evelyn Hutchinson (1958) used a similar metaphor when he expressed the niche concept as a multidimensional space. His concept of the niche began with a simple two-dimensional graph showing some character of the species against some property of its environment (fig. 23.1). By adding a third dimension, the graph shows a surface that describes the relationship among the three variables. We cannot add a fourth dimension on a two-dimensional page, but we can imagine it. Then we continue adding niche properties until we describe all the interacting factors required to define the species connections. Hutchinson concluded that the niche was a n-dimensional space— a niche hyperspace.

Coevolution and niche are concepts that focus on interactions among organisms and environment. Even though it is difficult to demonstrate genetic response in both members of the coevolutionary pair, it is not difficult to understand that individuals surviving and reproducing through positive interaction with an individual of another species could respond genetically and form a mutually supportive relationship. However, two coevolving species will not occupy the same niche; competitive exclusion prevents it.

Implications

The concepts of coevolution and niche are helpful in understanding how organisms come together and form interacting wholes. Coevolution looks at the relationship from the pair's perspective; niche considers the relationships from the individual species' perspective. The two concepts form a yin-yang pattern.

I have suggested that interactive relationships have evolved through natural selection. There is an advantage to predictable and ordered environments because energy and materials can be devoted to dealing with unexpected events, which may be the result of chaos. Relationship is a positive factor in an unpredictable world because it derives from the organism as a mechanism for survival. This means that interaction and relationship are not merely nice arrangements; they are essential to the survival of life.

But we must also ask, what are the costs of coevolution? Every benefit has costs. Clearly, a species locked into coevolutionary relationships is dependent on the survival of the other species. Independence is lost in coevolution. This is the basis for the metaphor I used to describe the widespread destruction caused by an environmental impact, where the event is like a stone thrown into a pond and the impact is like the widening circles

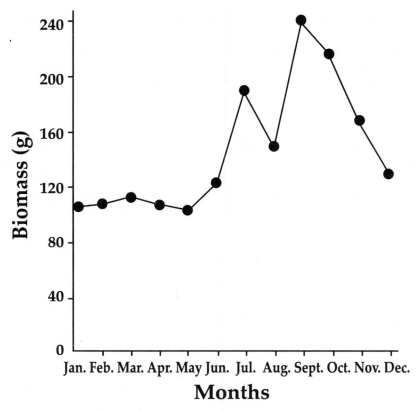

Fig. 23.1   A two-dimensional diagram of grass biomass and time in India.

of water around the stone. These circles sweep more and more species into the disturbance until the entire community is involved, often fatally. Knowing this, anyone who is sensitive to nature obviously tries to make as small impacts as possible.

An alternative argument about the nature of interaction in populations and communities is that chance plays the major role in directing who interacts with others. This argument focuses on the highly variable environment that affects organisms and moves them about, and on the highly mobile and actively creative response of organisms to their environment. This interpretation is also true. Some opportunistic species respond in this way. Probably many of them are the r species identified in Chapter 14. Chance plays a role in all ecological interactions because the physical environment is so variable and unpredictable and organisms are so responsive.

Once again we encounter a dualism in which order is contrasted with variation, which forms a circle of relationships. Our reasoning about natural systems might be better represented by the image of a Celtic brooch, with its complex intertwining of plants and animals, than by several lines crossing one another in an abstract space. These metaphors create yet another contrast emphasizing the complexity of nature.

Readings

Ehrlich, Paul R., and Peter H. Raven. 1964. "Butterflies and Plants: A Study in Co-evolution." *Evolution* 18: 586–608.

Futuyama, Douglas, and Montgomery Slatkin, eds. *Coevolution.* Sunderland, Mass., Sinauer Associates.

Pimentel, D., and F. A. Stone. 1968. "Evolution and Population Ecology of Parasite-Host Systems." *Canadian Entomologist* 100: 655–62.

24

The Biotic Community

The consideration of coevolution and the interaction of organisms for mutual benefit leads us back to the biotic community concept. The biotic community is defined as the entire collection of biological species living in a particular place at the same time. The community concept is one of the oldest concepts in ecology. Communities have been recognized from before the time of Ernst Haeckel in the mid-1800s to the present. For example, in the late 1800s Karl Mobius, a professor at Kiel, was studying the limits to oyster production off the German coast. He recognized that oysters lived in a community of organisms, consisting of both plants and animals. Mobius termed this collection a biocenosis, which means biological organisms living together. Ecologists have used many other technical words for the ecological community, but in English the simple word *community* has greatest currency.

Now that we have a large collection of ecological concepts, many of which deal with connections and relationships, we turn to the designing and constructing of a biotic community. First, let us address the question of whether such communities exist. Any country man or women will tell you how their landscape is organized into natural and built communities. Any field ecologist will identify a recognizable place where he or she has made a set of observations. These communities are not fixed; they change over time and have fuzzy boundaries. But history and experience tell us that there are indeed biotic communities.

Second, let us get rid of old-fashioned metaphors. Communities are not like individuals. They are not born and grow and die. They are not like human communities. Rather, our questions should focus on the structure, function, development, and relationship of biotic communities, because if we could understand these aspects, we could create a design and specify assembly rules for communities. At this time we do not know enough, but our objective is clear.

The Community Concept

The recognition of the community grows out of ecological experience with species interactions, such as competition, predation, and mutualism. For example, competitive relationships among species using similar resources, following Gause's concept of competitive exclusion, results in a division of the resource gradient among the competitors. Often one competitor has an advantage and assumes dominance, with the result that the abundances among species have the familiar shape of the species-area curve with a few common and many rare species. A common species may be a keystone species if its removal would cause the collapse of a food chain or a guild within the community. A keystone species such as white oak in an oak-hickory forest provides resources and services for many other members of the community that coevolved with it or depend on it.

An interesting example of a keystone species acting through predation to structure an ecological community comes from the work of Robert Paine of the University of Washington. Paine (1966) worked on the rocky tidal community along the coast of northern California and the Pacific Northwest. In this locality there are well-defined bands of marine organisms that live on rocks exposed to daily tidal and wave action. The communities consist of closely packed barnacles, mussels, starfish, algae, and other organisms that often fully cover the rock surface. Various seaweeds are present lower on the rocks, where they are more frequently submerged in the sea. Kelp beds and sea otters are present further out. Paine studied the abundance and the feeding relationships among the members of the rocky tidal community. He showed that barnacles and bivalves formed the base of the community, jointly occupying most of the space. These species were fed upon by *Thais*, a small gastropod mollusk, and all the species were fed upon by a carnivorous starfish, *Piaster*. These relationships are shown in figure 24.1.

What makes Paine's study interesting is that he then ran an experiment on a transect of the rocky intertidal habitat by removing the starfish from one transect and watching the changes in the abundance and spatial patterns of the prey organisms. The control area was not altered during the year-long experiment. A bivalve, *Mytilus californicus,* and two barnacles, *Balanus cariosus* and *Mitella polymerus,* formed a band across the rocks. The stable position of the barnacle and bivalve community was maintained by starfish predation. When the starfish was removed, the *Balanus* population increased and occupied 60–80 percent of the space, but six months later the *Balanus* were crowded out by *Mytilus* and *Mitella*. Successive re-

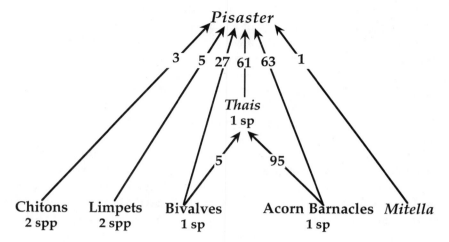

Fig. 24.1 Feeding relationships among species in a rocky tidal environment. The number of organisms consumed by predators is shown on the arrows. The number of *Piaster* was 1,049 and *Thais* was 287. From Paine 1966.

placement of species occurred and eventually the space was dominated by *Mytilus*. The benthic algae disappeared.

Removal of the starfish predator resulted in a loss of species diversity. The community changed from a fifteen-species to an eight-species system. The standing crop of animals increased. Further, several species not eaten by the starfish also disappeared, indicating that the starfish was indirectly influencing these organisms. Paine interpreted his experiment as showing that removal of the predator simplified the community, making it possible for one species to win at competition for space on the rock surface.

Mutualism

Egbert Leigh (1991), a scientist with the Smithsonian Tropical Biology Program in Panama, points out that in addition to predation and competition, mutualism structures communities. Individuals in small groups can use the tit for tat strategy, discussed in Chapter 20, to organize relationships. But in larger groups the strategy may be less effective because cheaters are encountered so infrequently that reciprocity isn't able to function as a control. In larger groups with multiple species, what may create a common interest that would suppress activity of those acting against the general welfare?

Leigh emphasizes the selective advantages of a balance among sexes, maintenance of the status quo, and stable neighbor and kin relationships in the evolution of mutualistic relationships. For example, in social insects, such as honey bees, the queen mates with several males and the sperm is mixed thoroughly. The eggs of the queen are brothers or sisters of the workers and are cared for by the workers. In contrast, eggs of workers are half-sisters or brothers to the workers and are eaten. The workers have a common interest in raising the eggs of the queen and maximize her reproduction. The division of labor in the colony is not under central control of the queen but is regulated by the age of the worker. The role of the worker is controlled by the titer of juvenile hormone in their body, which increases with age.

Leigh suggests that insect sociality might have evolved through mutual benefits, such as joint defense against predators, which favored group living. Group living may have intensified competition between individuals, leading to winners and losers. The losers might have adopted an alternative strategy that involved helping related winners reproduce, leading to a division of labor. Once subordinates were dependent on winners, loser reproduction could be suppressed, giving the whole group a common interest.

Species Redundancy

The diversity of species within a community is usually quite large. Hundreds or thousands of species may be present, but most are rare. The various guilds or subunits of the community are dominated by relatively few species, the keystone or the dominant species. What is the role of the rarer species? Do they make any essential contribution to the community, or are they merely there by chance?

This question is very important in conservation of biodiversity and, as a consequence, has received considerable attention. Rare species may be rare for many reasons. Some may be going extinct; they no longer can compete or have lost their genetic viability. Other rare species may be limited by the presence or absence of some factor in the environment. Remove or add that factor and the species would increase. Rare species may also be at the end of food chains. Relatively little energy is left to support their populations.

These ecological explanations of rareness can be contrasted to a statistical explanation, first presented by F. W. Preston (1948). Preston pointed out that the abundances of species in most communities fit a normal curve or frequency distribution. The normal curve is bell-shaped, with the two ends of the bell representing the extremes in abundance (fig. 24.2). The extremes represent rare or unusual conditions in comparison with the bulk of the observations that make up the body of the bell. Rareness in this case is merely an artifact of the statistical distribution of collections of species or objects of any type.

Which of these explanations is most likely in a given case? Essentially all species, rare or common, when studied carefully, are found to play a

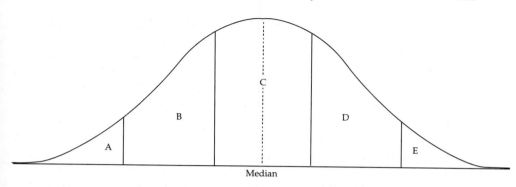

Fig. 24.2 A normal curve. *C* indicates the mean of the distribution. *B* and *D* are one standard deviation from the mean. *E* and *A* are two standard deviations from the mean.

role within a guild, food web, or community complex. Ecological research demonstrates that species fit into communities through feeding, interaction, and support roles. Recall David Schindler's experiment, where experimental treatment of pairs of lakes resulted in change in species dominance but maintenance of system function (Chapter 13). For this reason, contending that a community is merely a chance collection of species that happened to find their way to a location at a particular time seems counterintuitive. To think of nature as a giant game of chance may be appealing to some, but it doesn't accord with field research or actual experience. Yet we have to admit that experiments such as those of Paine and Schindler are difficult to carry out, and therefore the evidence of species roles in communities is limited.

Finally, the decision that species have roles to play in communities and guilds, and are not mere chance inhabitants of a locale, must be made in the context of evolution, dispersal, interaction, and the other phenomena governing species development, survival, and extinction. A community has fuzzy boundaries in space-time, the components of guilds are dynamic and ever-changing, and the species are also dynamic. Nevertheless, individuals and species live within complex biological units, including communities, are subject to the physical-chemical environment, and interact with other individuals and species to cope with the environment. Those that survive reproduce and contribute genes to the next generation. Such communities inhabited by essentially the same species in the same relationships are scattered across the landscape. Chance plays a role in this dynamism, of course; but it is merely one of the phenomena shaping the community.

Implications

At this point it is enough to note what this discussion of community organization means to the environmentalist. Species, individuals, and habitats are unique. It is not possible to generalize across species, individuals, and habitats without loss of their particular character, which may be significant in their survival and role. This means that we must value place and organism in their localness and in their specialness, but it does not imply that we *cannot* generalize across species, individuals, and habitats. We can make generalizations, of course, but in doing so we shift the emphasis from the organism and place to an abstraction of the human mind. If we carry this very far, we have mathematical models that reflect only a few quantitative properties of the original subject. This abstraction may be useful for reasoning, but it is not a suitable base upon which to describe or manage nature.

This conclusion is very unsatisfactory to natural resource managers or environmental bureaucrats who are facing complex political, social, and economic pressures and would like nature to be simple. They are tempted to turn aside from the ecologist who honestly says, "I do not know the assembly rules of a biotic community" to someone else who, for professional or personal reasons, claims to have the answers. If the experts or managers of nature pay no costs for their errors, why bother to know? If our life style allows us to avoid any contact with nature, why be concerned about it? If we can transfer costs of management to future generations or other nations, why not do so and avoid today's costs? How does one face angry citizens who claim to need water to maintain cultural icons such as green lawns in a summer drought and say that we cannot pump more water because we have to maintain river depth to preserve the fish fauna? No one should underestimate the complexity and difficulty of the natural resource manager's job in dealing with an explosive and greedy human population.

Eventually, of course, the rain returns and citizens ask why the crisis mentality was necessary. The river is full again, fish have survived (maybe in smaller numbers) in some refuge, and life has returned to "normal." We could have dried the river up and it wouldn't have mattered. Aldo Leopold said that to be a conservationist is to live in a world of wounds. This is certainly true.

Readings

Drake, J. A. 1990. "Communities as Assembled Structures: Do Rules Govern Pattern?" *Tree* 5: 159–64.

Roughgarden, J. 1989. "The Structure and Assemblage of Communities," pp. 293–326. In J. Roughgarden, R. M. May, and S. A. Levin, eds., *Perspectives in Ecological Theory*. Princeton, N.J., Princeton University Press.

Whittaker, R. H. 1962. "Classification of Natural Communities." *Botanical Review* 28: 1–239.

25

Island Biogeography

Communities can form in any medium. If the location is a mature terrestrial landscape, the physical-chemical environment may be relatively homogeneous and constant over time. If it is a marine continental shelf, then the habitat for organisms may be highly dynamic, changing physically and chemically by the hour or day. The opportunity for communities to form under these ecological conditions will be different. But no matter what the local conditions may be, each will have a dynamically changing species composition, because species are continually arriving and leaving a community. Not only are the species within a community evolving and adapting, but the species composition is changing through dispersal. It is difficult to study this process in these terrestrial or marine cases because the boundaries of the communities are so fuzzy and the connections to other communities so many. It is much easier to study dispersal and extinction of species on islands.

An island conveys a romantic image, signifying isolation, freedom from disturbing life, a place to get away, a vacation. Islands have these images because they are separated from the mainland and are therefore isolated to some degree. Because of this separation, they also tend to have special ecological characteristics. Their biota is restricted to organisms capable of passing over the sea to reach the shore and then survive and multiply. As a consequence, some islands have rare and unusual species that are found nowhere else. These special features of islands have attracted field ecologists, along with the vacationers and those seeking escape.

Island biogeography is more than a curious subject of remote places. Quasi-islands are common in the terrestrial environment, and we can extend some of the conclusions derived from study of oceanic islands to these isolated communities in the landscape. Terrestrial islands might be mountaintops surrounded by inhospitable grassland or forest. Humans also create islands, and any glance at a humanized landscape reveals an archipelago of islands of forest fragments or fields in a matrix of another land-use type. As a consequence, island biogeography has become a topic of interest to conservationists and landscape architects, who use biogeographic theory in the design and management of natural reserves. Besides these special features of island biogeography, they also provide an insight into the organization of ecological communities in general.

Islands and Their Biota

Island biogeography involves the interaction of the biota with an isolated habitat. In order to understand the interaction, it is necessary to examine both the biological organisms that may disperse to the island and then occupy it and the island habitat, its hospitality to organisms, and the nature of the barriers between it and source habitats.

The rate at which species reach an island depends on its distance from the source of the organisms. Distant islands receive species at a much slower rate than those near the mainland. Of course, the rate of dispersal depends on the kind of organism that is a potential disperser. Land mammals tend to disperse slowly to oceanic islands, while plants and birds disperse much more rapidly because they can move over the water to the island. Insects also can move relatively long distances over hostile habitat, because they can float on water or ride air currents and winds. Higher mammals, reptiles, and nonflying, nonswimming life forms need rafts or some other device to transport them. Many organisms have hitched a ride with humans. The consequence of the problem of distance is that the number and diversity of species decline with the degree of isolation of the island.

Once a species arrives on an island, it must establish itself in a suitable habitat. Successful establishment requires an adequate physical environment but also, assuming that other species are already present, competitive ability, so that the new arrival can find a suitable place. The ultimate number of species accommodated on an island is a function of its size and the complexity of its habitats. For example, the relationship between the diversity of amphibians and reptiles on islands in the West Indies and island size forms a straight line, with a slope of 0.3 (fig. 25.1). The species-area relationship on mainland habitats is much lower. The curve describing the diversity of the North American bird fauna and size of area between ten acres and one million acres has a slope of about 0.17 (fig. 25.2). Note, however, that the curve for North American birds is not a straight line—for very small or large areas, the slope of the line is much greater. The apparent rea-

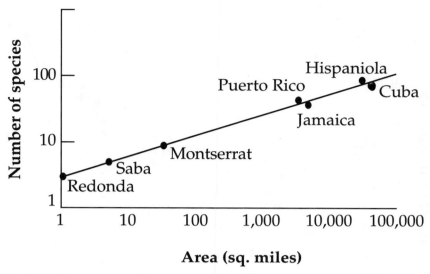

Fig. 25.1 The species-area curve for amphibians and reptiles on islands of the West Indies. From MacArthur and Wilson 1967.

son for the difference in the species-area curve on the mainland and on islands is that many transients cross the boundaries between units of different scale on the mainland, especially birds, which are highly mobile. Transients are less frequently encountered on islands.

These patterns are different for islands in the terrestrial landscape. On mountaintops, bird and mammal diversity has a different relationship with area. Mammals on the mountaintop are usually remnants of a once widely dispersed fauna and are now relict populations. Unless corridors for mammal movement have been constructed between mountaintop islands, the mammal fauna has little chance of receiving new dispersers. In contrast, birds move between islands and maintain higher levels of diversity.

Equilibrium

On a particular island the diversity of the biota depends on the input of species arriving by immigration and on the loss of species through extinction. The input rate, based on successful immigration and establishment, declines as the island biota is filled with earlier arrivals (fig. 25.3) and falls to zero when all the species that potentially could be present (representing the species pool) are present. The loss rate due to extinction increases as the island becomes more crowded with species, due to competitive interactions for resources. The point at which these curves of immigration and

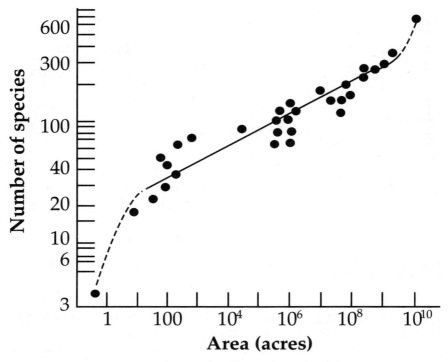

Fig. 25.2 The species-area curve for North American birds. From Krebs 1985.

extinction cross indicates an equilibrium toward which the biota of the island will move.

This model is very simple and does not include any information about the quality of the environment. When we add the size of habitat to the model, we obtain a family of lines illustrating various hypothetical patterns (fig. 25.4). The extinction line is depressed for larger islands. Small islands have much higher rates of extinction than large islands. The distance to the mainland causes variable effects on immigration. The immigration rate is higher the closer the island is to the mainland. Altering these factors creates a variety of equilibrium points.

Biogeographers caution that these theoretical ideas, which were first presented by R. H. MacArthur and E. O. Wilson (1967), cannot be applied without understanding that the patterns depend on many other factors associated with the biota and the habitat. Each particular case requires study and analysis. For example, on some islands the geological history involved with plate tectonics has influenced the isolation and diversity of habitats.

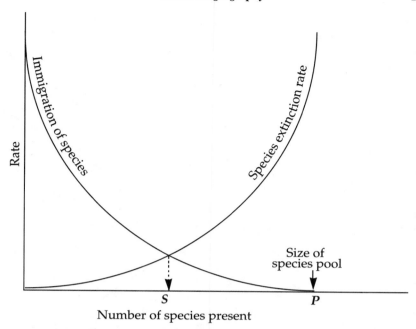

Fig. 25.3 An equilibrium model of a biota on a single island. The equilibrium species number (S) is reached at the intersection of the curve of the rate of immigration of new species not already on the island and the rate of extinction of species from the island. From MacArthur and Wilson 1967.

Islands that were connected to the mainland during the Pleistocene when the sea level was lower became isolated relatively recently and may still be moving toward equilibrium. In this case, extinction of the original mainland biota may still be occurring, as the biota moves toward the species diversity that fits an island of that particular size and habitat variety. New islands formed by volcanic action are being invaded and the biota is slowly increasing.

Where islands are being formed on land, as when forests are fragmented by logging and development, the situation is also complicated. Extinction occurs because the biota in the fragment represents an arbitrary selection of species that happened to be present when the island was formed. In some cases insufficient numbers of organisms remain to maintain a population; in other cases the resources are inadequate to sustain the species. The fragments also contain a new kind of habitat—an edge environment—which is a mixture of conditions of the fragment and the new matrix of agricultural fields or the cut-over land in which the fragment sits. A very small fragment might be made up entirely of edge habitat. Species characteristic of edge environments occupy these new areas and further reduce the resources in

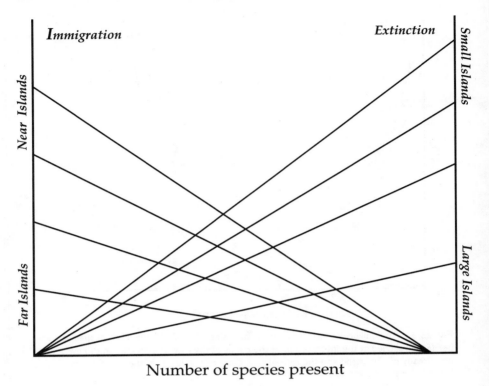

Number of species present

Fig. 25.4 Equilibrium models of biotas on islands of different size located at varying distances from the source area. From MacArthur and Wilson 1967.

the fragment, as well as parasitizing and competing with the resident biota. To maintain the resident breeding birds in a temperate forest landscape may require a landscape patch of a few hectares. But to maintain large predatory mammals and birds the island might need to be very large.

Clearly, the caution of biogeographers is warranted. Nevertheless, the presence of islands and the high rate of fragmentation of habitat by human disturbance make the concept of island biogeography relevant and practical in natural resource management. The theory suggests that we may want to enhance the opportunity for migration by maintaining corridors between fragments. We may also limit extinction rates by conserving fragments as large as possible and fragments with considerable habitat diversity. The object of management is to move the equilibrium point toward the maximum number of species possible.

Readings

Abbott, I. 1980. "Theories Dealing with Land Birds on Islands. Advances in Ecological Research 11: 329–71.

Gilberet, F. 1980. "The Equilibrium Theory of Biogeography: Fact or Fiction?" Journal of Biogeography 7:209–35.

Haila, Yrjo. 1986. "On the Semiotic Dimension of Ecological Theory: The Case of Island Biogeography." *Biology and Philosophy* 1: 377–87.

26

Human Ecology

Throughout this book I have assumed that humans are subject to the same ecological controls, pathways, and opportunities as other organisms. That is, I have assumed that humans are part of nature, and that there is no divide between humans and the natural world. Not everyone would agree with this assumption. The nature of the relationship between humans and nature is one of the deep philosophical questions of our culture. Developing an answer to this question is far beyond the boundaries of this book.

There is, however, another question involving the application of environmental concepts to human beings that we should focus on. This question is methodological: How can we distinguish the effects of environmental factors on humans from all the other influences humans are subject to? Recall that ecological systems are characterized by complexity, by continual change in response to environmental influences, by internal adjustment to selective forces, and so on. How can we tease apart the causation in such systems and assign cause to one or another factor among hundreds?

Although we can fit human beings into our hierarchical structure and treat the human as an individual, a population, a metapopulation, or a species, *Homo sapiens*, humans have many special attributes that make treating them as a whole difficult. An example of this problem is found in medical practice. Seriously ill people are placed in an intensive-care facility where the environment is tightly controlled. Under these conditions they tend to act as laboratory animals, responding to treatment in a predictable way, with exceptions due to their genetic constitution, past experience, type of injury, and so forth. The point is that the high degree of control permits the specialists to focus drugs, treatment, and surgery on the problem and hold the organism in a physiologically optimal condition. Now let us assume that the individual has begun to recover and after a period under more conventional hospital care is sent home. In the home environment some people recover faster than in the hospital. They benefit from a familiar environment with family members as care givers. Others suffer from a negative home environment and may get worse. Here is an example where change in the environment, defined as a complex of interacting factors in the home, can be positive or negative given the individual, the type of injury, and the home experience. But what are the specific factors that influence recovery? It is very difficult to say.

The problem of assigning cause has been present throughout my discussion of most of the basic environmental concepts. But it is especially acute in human ecology, because humans have several special traits that are integrative and that defy analysis or division into parts. Integrative traits include the development and use of language and signs, the presence of culture, and the creation of technology. There are other characteristics, but these will do to make the point clear.

Language

The ability to think abstractly, to create new ideas or objects, and to form social groups depends on language. Without speech, human mental ability would be much less effective and much less flexible. The brain would be a blunt rather than a sharp tool.

Language is a universal property of humans. Other organisms communicate with each other by sound and signals, but no other animal has the complex language forms of mankind. Complex language is found even among people with little technology. Language also changes continually. It serves our purpose of communication and self-expression, and if it is insufficient we make up words and expressions and grammars to do the job. The young especially manipulate language, and it seems as if each generation creates its own superlatives, expletives, and codes.

Language also divides humans: witness not only the familiar difference between language groups but also misunderstandings between individuals. All of us find it impossible to convey to others what is in our mind exactly as we experience it. And we are frustrated about how often our intentions are misunderstood by family and friends, who should be able to understand us. The message that we send in a shared language is decoded by the other in a way entirely different from our intention. These errors in communication are a serious barrier to joint purpose.

Careful study shows that languages organize human social groups. There are languages of trades and occupations, of gender, of age, of economic condition, and of race and caste. Every division among humans is fixed in language. This means that one way humans structure society is through the language they speak. The size of the group must not exceed the capacity to speak from individual to individual, yet the group must be open enough to share language with other groups as necessary. Frequently errors in environmental management and understanding can be attributed to problems with language. Words mean different things to people. The way new ideas are expressed can be threatening. Unfamiliarity with the polite words and the proper words can elicit a negative response from the listener.

Culture

Humans have a collective sense of the social whole, which we term *culture*. Culture involves how we think, how we speak, how we act, and what we do. It has a similar sense to the term *niche* as applied to an animal or plant species. Culture is learned from birth and is reinforced by ritual, ceremony, social activities, and conventions. Culture instructs us on what to do when we encounter an unfamiliar situation. Individual creativity works within culture. That which is outside of culture is either ignored, feared, or destroyed.

All human groups have culture, but culture varies to the same degree that language varies. Culture appears to be an expression of human adaptation to complex environmental conditions. Yet human adaptation is changing as environmental conditions change, and so we frequently find that cultures contain a variety of adaptive features which probably arose as a response to environmental conditions in the past. Since cultures are conservative, adaptive traits might be carried beyond the point where they are useful. As a consequence, cultures are interesting attics, with old trunks full of no longer useful but still interesting relics, along with the newest inventions.

There is no way to escape culture. It is as tightly bound to us as is environment. We are penetrated by culture; our actions and thoughts are shaped by it. Yet we are unconscious of culture most of the time. Only when we have the opportunity to live in another culture for a long time do we see our own culture as different when we return to it. At that point the peculiar ways in which we think and act are visible. But soon our culture disappears from our view and we treat it as the right or commonsensical way of doing things.

Although the human individual can think and act creatively, expressing individuality in a unique way, the creation must be expressed through culture and accepted within a culture. Otherwise it is rejected. This is the source of the problem of creative intellectuals or artists who find their personal expressions misunderstood. Culture is a form of adaptation, and although it changes, human groups tend not to risk what has worked in the past. We tend, collectively, to be conservatives.

This is why environmentalists must learn to manage cultural change. In order to alter our consumptive and destructive patterns of living, we need a different cultural model that stresses maintenance, reuse, cycling of materials, cooperation, mutualism, and the other properties discussed in this book. These concepts must be taught to children by parents and schools. We must have examples of successful adaptations before our eyes so that we can imitate and improve on them. We must have laws and regulations that move us toward positive adaptation. Eventually these will create a changed culture.

Technology

Finally, we can consider the ability of humans to create technology and, through technology, the built environment. This trait is grounded in a general property of organisms. Birds construct nests, bees construct elaborate hives, beavers construct dams and reshape the landscape. Thus, like language, humans are located at the end of a continuum within the animal kingdom. Only humans have produced a technology that has permitted them to create a new kind of environment and reshape humanness itself. Application of this trait globally now threatens many other species.

The city is an example of a built environment. Cities do not occur in nature. They were invented by humans several thousand years ago. The United Nations predicts that more than half of all humans will live in cities by early in the next century. Probably most urban humans never encounter nature uncontrolled by humans—that is, wild nature or wilderness. But the fact that 50 percent of humans will not live in cities does not mean that they will have an appreciation for wild nature either. Almost all of the other half of the human race lives in another form of built environment that we call the agricultural landscape. As we can see almost anywhere in America, Asia, and Europe, all the land in agricultural landscapes is shaped by human activity. Humans have created a new rural and urban world through technology and culture. There is no need to adapt to nature, because nature does not exist as a coherent system in the human-dominated technosystem. Rather, it exists as populations and as individuals who cause problems. The image of a child terrified by a "bug" is a commonplace in modern societies. Urban humans often are frightened of or at least uncomfortable in nature.

But if technology is part of the ecology of the human species, can it be controlled or directed so that it is positive and increases human well-being? I think that we must answer yes to this question, without having much evidence for our answer. Technologies grow from human creativity, which is usually but not always focused on solving problems identified by humans. Problems are recognized within certain forms of thought and certain life styles. It requires a specific set of circumstances for a problem to emerge. Thus, the initiating cause and the form of the response are not fixed at all. But they are accessible to reason and to ethical rules. We can, for example, treat wastes in holding ponds and in waste treatment facilities instead of dumping them in the nearest river. But to do so requires an ability to view rivers as transportation networks of landscapes and to respect the rights of others to use unpolluted river water. The result of these attitudes is a willingness to accept the additional cost of waste treatment because it is right to do so. Unwillingness to accept the cost may be based on ignorance—that is, on not knowing that we are interconnected through rivers. Or it may be based on power—on not caring what happens to others. Or it may be based on economics—on the belief that by saving the cost of waste cleanup and transferring the cost to someone else, one is ahead in the competitive game.

It is hard to imagine how human attitudes might be changed to incorporate limits to the destruction of nature into human society. Frederick Ferré (1988), who has studied the philosophy of technology, suggests that it might come from applying ecological concepts more directly to human decisions. I agree. That is one reason for writing this small book. It is a hopeful idea, but I also have suggested in the previous pages various limitations to its accomplishment.

Human Ecology

To sum up, factors such as language, culture, and technology complicate the analysis of human ecology. For this reason many human ecology studies describe what is observed and do not attempt to explain why the observed structure and function evolved or developed that way. Frequently, an understanding of such phenomena as culture, technology, and language comes from long experience and intelligent observation. There is a gestalt sense of the whole, which we can express in a systems context. Either way, it is essential to link environmental concept and human experience to fit human life within nature.

Readings

Cronan, William 1988. *Changes in the Land: Indians, Colonists and the Ecology of New England.* New York, Hill and Wang.

Ferré, Frederick. 1988. *Philosophy of Technology.* Englewood Cliffs, N.J., Prentice Hall.

Young, Gerald L. 1989. "A Conceptual Framework for an Interdisciplinary Human Ecology." *Acta Oecologiae Hominis* no. 1, University of Lund, Sweden.

Conclusion:
Ecology, Environment, and Ethics

In this chapter my objective is to summarize the key elements of the technical concepts presented earlier and then to reason from them toward an environmental attitude and practice that derives from environmental science. I say "derived from environmental science" on purpose. Practical work can be derived from or founded on scientific findings and understanding, but science is limited by its point of view and its method. It seems odd to claim that environmental science is narrow, when it studies hierarchies that range from the planet to individual organisms. But the environmental sciences often do not include people in their studies, and almost never economics and politics. If they exclude these, how can they be practical?

Four features characterize environmental concepts: dynamism, connectedness, creativity, and limitations.

Dynamism

Ecological systems are dynamic and vary over space and time. We search for and find generality in nature because our minds work that way. We need the general in order not to spend all our time focusing on the particular. For this reason, I find no philosophical problem in the contrast between the particular and the general in ecology. Nature is made up of difference, but we arrange nature into collections of like objects and processes for our convenience. In doing so we pay a cost. The cost involves overlooking real differences between members of the collection. If these real differences are important, then our collection may be false, and if we insist upon it, we will have to pay the costs of error.

I have proposed that all natural processes, including those of the human-built environment, are dynamic in space-time. Some, such as a mountain or a river, seem fixed, but that conclusion is an artifact of our life span and visual apparatus. When a vast river system floods or mountains explode or break apart, we see the dynamic behavior of these systems as well.

This underlying idea means that all is change—life and nature. Although humans may counter this dynamism with the built environment and machines for living, these constructions also change physically in time, and our attitude toward them and our use of them also change. A human life focused on the life history of humanness and the dynamics of nature is often much happier than one focused on retaining into the future some specific age or condition of life.

Dynamism is derived from the movements of the Earth surface, the movement of the atmosphere and hydrosphere driven by the Earth's energy balance and evolution and natural selection of the biota. The ecologist's task is to trace some of the ceaseless movement and to understand its causation. For this reason environmental work focuses on a habitat, a biome, or a group of organisms, with the goal of becoming intimately knowledgeable about them. There is no other way to build the information and understanding of nature that enables one to see patterns.

The patterns are knowable in different ways. Large, slow patterns are difficult to see because of their size and our short life span and comparative smallness. To see the small, fast actions requires special instruments. We have been enormously successful in creating instruments to look into this aspect of nature, so that now we can even look inside the atom and see quarks. We have been less successful in creating instruments to see at large scales. Probably the space efforts of several governments have been most helpful in giving us a global viewpoint. Human limitations cause us to focus on medium-scale events—the scale in which we live. The partly educated citizen often finds large- or small-scale events to be threatening.

Dynamism requires that we give up the notion that the world and life are static and fixed. No objects in nature are unchanging. If someone declares that such and such an event or object is unchanging, it is because we have defined them in such a way that we can ignore them. Of course, this strategy lasts only so long, and then we are surprised when dynamism forces itself on our consciousness. A simple illustration is the great floods of 1993 in the Mississippi River system of the United States. This system is so large that continent-scale weather patterns influence its dynamics. The public, ignorant of hydrology and taking a static approach to landscapes, demanded that the river be held within certain limits by levees and dams. The engineers bragged that they had brought the Mississippi under "control." The potholes and wetlands in the prairie watersheds were drained and converted to farmland, making it impossible for them to absorb water. The flood plains too were converted into farmland and towns. All of this development was threatened during a period of prolonged rainfall. The levees were inadequate, the river flooded across them, farmland was ruined, towns were evacuated, and the country paid the price of a static public policy.

Newspaper reports suggest that the people did not learn from this experience. They have returned to their farms and towns and demanded larger levees and more flood controls. The question is, who pays for these developments? If the people living in the Mississippi flood plain had to pay the costs themselves, derived from the farmland and the commercial activities of the towns and communities of the region, they would be forced to adapt to the natural dynamism of the river system. But if the costs are distributed countrywide, those in power are free to satisfy the political demands of the injured floodplain dwellers, the taxpayers submit to a small and virtually unnoticed extra tax, and the region, having failed to adapt, waits for the next catastrophic rainfall.

Eventually, even a rich country such as the United States will become exhausted from working against nature and will be unable to maintain this effort. But a poor country, such as Bangladesh, just accepts the environmental and social losses and goes on. The solution to this problem is to work with nature, to take advantage of channels and wetlands which are required only during the rare storm and recognize the multiple values of floodplains and wetlands. An excellent example of a successful restoration of a smaller river system is the development of the Desplains River north of Chicago. Here the floodplain was reestablished and the natural functions of the river were restored by ecological engineering techniques. As a result, the floods were no longer destructive, wildlife returned, and the river margins became popular recreational areas.

Connectedness

The fundamental theme of this book has been connectedness. Ecology is the study of connections, so the theme is appropriate for the subject. I have presented the theme in a systems context. This approach has the advantage of being familiar to the modern reader, but it has the disadvantage of reinforcing mechanical and abstract metaphors. How does one speak about connection in a culture of separation and isolation? I don't know. I do not feel that I have completely solved the problem.

The sense of connectedness is difficult to sustain consciously. We live continually in environments that interact with us on every level through which interaction can occur. When we focus on some aspect of these environments, the other parts fade from our attention. It is as if there is a gradient of visibility that fades in both directions from the center of our momentary interest. We shift our attention from one part to another, but we cannot easily sense the whole.

The capacity to sense wholeness is a highly valued trait in certain cultures. It often comes to us from long, intense experience with place. Patterns become familiar, and when they change one senses their presence. But this knowledge is usually at a level that cannot be communicated; it is merely perceived. There are other ways to obtain a sense of wholeness. Often they involve disciplined training in meditation. If I understand it correctly, students are taught how to avoid focusing on particular aspects of their environment and how to become integrated with it as a whole. Meditation in Zen Buddhism is an example of this type of exercise.

Connectedness requires a response from us in two ways. First, we need to acknowledge the connections we make with nature and the other. When students ask me how to do this, my answer is prayer. Prayer is a private, silent statement recognizing a connection. We need to express our understanding of connectedness by consciously affirming that we are part of environments, connected to them in ways we do not understand, and that we accept a responsibility for being connected to another. A simple statement of the fact of being connected is a first step toward a deeper acknowledgment of one's role and responsibility. Then, in time, the prayer can become more personal and richer with intention and meaning. Second, there is the need to be actively engaged in playing the human role in the environment. Being human is a great responsibility. It involves knowing about nature. It means that we can express our knowledge in spiritual terms. It means that we can understand responsibility. The environment needs humans. Humans enhance the environment. The challenge is to convert humanness from being the destroyer of nature and productivity to being the supporter of nature, the sensitive gardener, the caring mother or husbandman.

It is difficult to focus on connectivity in a culture that emphasizes separation. Yet it does not help to isolate oneself further by withdrawing from the culture. Just as there is no "away" to receive our waste, there is no "away" to go to and escape the culture. Rather, we should recognize that dynamic behavior means that there is always a positive and a negative, a cost and a benefit, a producer and a receiver in every contrast we construct mentally. Cultural negativity can be converted by focusing on the positive aspects of dynamism and using them to create solutions to problems and to support nature. The task is to increase life-enhancing activities and reduce life-destroying activities in the culture, to reduce the speed and dimension of human behavior so that they fit the natural scale, and to recognize the limits of environmental dynamism.

Dynamic action results in a myriad of interactions that create cycles of interactions. It has been difficult to emphasize these cycles except through words. Cycles are everywhere; all is circular. What this means to the individual is that it is impossible to isolate one action. Every action undergoes a widening as reactions begin to build and then connect, forming new responses. In this context, human responsibility, although unique in nature, becomes of enormous significance. We are connected and we recognize our connectedness and we have responsibility for the connections that link us to the complex world. There is no escape. This is one way to interpret the creation story of the Christian Bible. We are uniquely endowed in the creation, and we are directed to use our unique powers as stewards of the creation. To destroy or misuse the creation is blasphemy.

Creativity

The third aspect of the ecological story grows naturally out of my comments on connectedness. Creativity involves the directionality of the dynamic process. Dynamism is not mere randomness. It is not change for change's sake. It is not chaos. Rather, dynamism operates within specific contexts, which order and control it. For example, biological evolution involves change in genetic structure, which produces phenotypic variation, which is selected through interaction of the biota and the environment. The result of evolution is that organisms adapt to environmental change. The creativity is such that often the adjustment is unnoticeable, except to the watching ecologist and geneticist. But the adjustment also may be inadequate and the organisms may die, or it may burst out into a quite new adaptation that is noticeable to us. The directionality involves the close coupling with the environment. But environment is not a thing; it is a complex of organisms and physical-chemical processes all changing in their own ways. Creativity always takes place within this system of interactions. Fitness involves being fit to the entire system of interactions, not to one or another environmental limiting factors.

Recognition of creativity in the natural world is not commonplace in our culture. If we notice it, we interpret it as natural, as part of what should happen. In contrast, we value human creativity highly. It is associated with a creator, whom we honor and endow with special qualities. When the creative act is unique and commands a market value, we consider it most highly. The farmer who overcomes poor weather and makes a crop, or the businessman who overcomes late delivery of stock and weak demand to turn a profit, we do not consider creative in the same sense as the painter or poet. A plan for future development of a city, recommending how we should use land and water and arrange houses and streets, does not usually qualify as a creative product. Our cultural interpretation of the nature of human creativity, then, is ambiguous. It seems to require an individual creator as well as a product of creation, and the product needs to have a value that enhances the status of the owner. Creativity and power are related in Euro-American culture.

Natural creativity is different. It involves adaptation to the environment, adjusting to outside forces with the resources available. It may be so subtle that we do not notice it. It is not attached to an individual, nor to value nor power. The value that it represents is value for the whole. It involves fitting into the whole and not appearing as a single, separated creative act. One would think that something that has produced us, that endows the Earth with a unique physical chemistry, that produced and maintains the complex biodiversity in which we live, would be highly valued too. But it is not, and indeed there are those who deny that natural creativity exists.

Limitation

Those searching for answers to environmental problems often are disappointed by ecologists. When placed in an either-or situation—for example, on a witness stand—ecologists tend to say that they are unsure of the consequences of an action. They point out that consequences depend on many factors, and that internal variation makes a deterministic outcome unlikely. The ecologist's uncertainty is frustrating to the engineer and the lawyer who are quick to say confidently that this and this will happen. Why are these specialists different? How shall we evaluate their performance?

Engineering is applied physical science, in large part, and it deals with processes that are more deterministic than ecological processes. When we place two chemicals together, the result is expected; when we place two chemicals together in an environment containing a variety of microorganisms, the result is much less predictable. Further, the engineer restricts his or her decision to the physical process and leaves the "social issues" to the decisionmaker. The ecologist looks at the whole system, which includes social issues. The engineer excludes from the model processes that are poorly understood or highly variable and takes care of them through a safety or contingency factor. Ecologists try to understand the interaction of all the parts of the process and hold themselves to relatively high levels of statistical certainty.

Clearly, there are substantial differences between ecology and disciplines that take a more narrow and focused view of their objectives, such as engineering. Many ecologists urge their fellows to imitate these disciplines and become more influential as a consequence. Yet ecologists work in the real world, where they encounter the concepts described in this book. The deep divide between the natural world of dynamic variation and connectivity and the human-conceived world of linear process and separation is real. There truly are several points of view.

It is important to be clear that these points of view cannot be contrasted and declared right or wrong. They exist, are relevant, and must interact. The deterministic approach is valuable and essential when we are dealing with a process that can be fairly clearly isolated from other processes. Many processes in the built environment have this character, and for these the deterministic approach can be successful. When we deal with the social or natural environment, the situation is different. Our ability to predict is much less, our knowledge is less, and the costs of our mistakes may be larger and more enduring. In this case, careful, considerate, and sensitive analysis and management are required. This approach is the opposite of the gung-ho, let's-get-it-done attitude of the builder who is working from a well-known plan with specified materials.

But it would be misleading to say that ecology does not have the same confidence in its principles as engineering. Certainly, many of the subjects treated in this book are well understood. Hydrologic processes, biogeochemistry, the physiological basis of primary productivity, population growth, and competition are well understood, and the ecologist usually feels confident in reasoning from what he or she knows to a new application. But in other cases we are less certain. As always, everything depends on the circumstance, the question, the time, the place. There are no generalities that cover all situations.

Applications

The overarching idea of ecological concepts is that the biosphere is a web of dynamic interactions. But the presence of an interaction does not tell us about the pattern or rate of flow from one to the other side, between the natural and the human. The final question is, is there a limit to human use of nature to satisfy human demand? Since humans are rational animals and can imagine the future, do we have an obligation to act differently from other organisms and limit and control interaction from our side of the interaction network?

The answer to this question has many aspects and must be deconstructed to obtain a useful answer. First, it is clear that humans are enormously successful animals with the ability to adapt to a variety of environmental conditions. Like any organism, humans have vital needs and it is appropriate that these needs be satisfied from nature. However, the definition of "vital" is culturally determined. Modern Western culture defines it to include a large material life and an energy-demanding life style. Other cultures have a less rich material life. Apparently, there is no limit to the capacity of humans to consume material goods and services. Thus, it is unclear what vital needs are.

Second, the capacity of ecological systems to withstand abuse by human demand is not clearly understood. Where physical forces destroy ecosystems, as on Mount St. Helens in Washington State, biological life slowly recovers on the eroded and blasted surfaces. In small, shifting agricultural plots in the tropics, recovery can be quite rapid, but in regions where large areas of forests have been cleared, soil erosion, leaching of nutrients, loss of seeds in the soil seed bank, and loss of native animals who disperse seeds all create a situation that is disastrous from any point of view. These landscapes produce a crop for two or three years, then are converted to a grassland that supports few cattle. The grasslands burn, destroying forest reproduction. Ultimately, the rich forest landscape is turned into an unproductive grassland with no economic value and few natural values. However, in cultures where social power is correlated with the amount of land owned by an individual, owning large acreages of unproductive grassland is sufficiently attractive to perpetuate the destructive practices.

The resiliency and resistance of many systems are poorly understood. Some ecologists, reasoning from machine analogies, have claimed that an ecosystem will collapse after intense abuse. They cite as evidence the collapse of a population, such as the cod fisheries in the north Atlantic, the growth of an aggressive plant, such as blue-green algae in polluted ponds, or the destruction of tropical forest landscape. In these instances ecosystem collapse is actually rapid change due to the decline or growth of a species or a group of species, or the loss of the productive capacity of the land due to intense heat and rain. We don't know if an ecological system will fall apart when a particular number of species are removed, or a specific quantity of biomass is destroyed, or a given quantity of essential elements is leached from the system. It is not clear what the word *collapse* means in these situations.

Third, if we have vital needs and cannot appeal to a clear and concise measure of the capacity of ecosystems to withstand human demand, then the control of demand must come from within humans. Our willingness to curb our demands on behalf of those on the other side of the interaction depends in part on our shared sense of community with them. If the other is kin, we are more likely to share or limit our demand than if the other is a stranger. This means that if we can build a relationship with nature, with wild organisms and natural processes, we are more likely to limit our demand, if it can be shown that our demand is destructive. George Williams, in his attack on the Gaia Hypothesis (1992), declares the environment to be hostile to humans. I think that he is wrong. I find the environment to be comfortable, welcoming, and supportive of humans. Clearly, we must give more people a positive experience with nature so that they see it as worthy of their moral consideration.

Fourth, the way we shape and limit our demand on nature varies with culture and circumstance. If we can extend our concept of neighbor to include our local environment, then we might extend the golden rule, Love thy neighbor as thyself, to the environment. How do we express love to the environment? Love refers to a deeply emotional human relationship. Can it be extended to nature? I doubt it, although I have acquaintances who do express their feeling about nature in that way. But we can respect nature and its right to function uncontrolled by humans even if we cannot love it. We can rediscover how carefully to divide the land and water among natural and humanized landscapes. Eugene Odum declared that approximately half the landscape should be left to operate according to natural processes in order to provide the humanized landscape with high-quality goods and services. If we agree, the question then becomes, where do we locate natural and humanized landscape elements and how do we shape their interactions? To answer this question is one of the tasks of landscape ecology.

Certainly, meeting vital needs does not include foolishness, such as harvesting and selling natural products beneath their production cost to satisfy the demands of political pressure, providing services at less than market cost and charging the taxpayer for the difference, or destroying the environment out of anger at environmentalists.

Finally, Paul Shepard's (1982) comment that nature is vital to human sanity seems to me to focus on the key point of the entire discussion. Without nature, humans have no other, no creation, no unknowable to use to shape humanness. Without nature, human desire and dreams becomes all there is, and who is to judge which dream or fantasy is realistic, better, or wrong? The human spirit then can drift into any nightmare imaginable and the nightmare will be declared normal by those who have power over it. The past century offers much evidence of social insanity designed and maintained by intelligent, good people. It is a special weakness of the technocrat. Humans, especially young humans passing from childhood to adulthood, need the challenge of the unknown to shape the self. Otherwise, one only looks in a mirror and sees the human face. Nature is not a mirror. It operates on its own rules, which change in space-time. As a consequence, the human is put on his or her mettle. And one begins to learn about one's abilities and limits and to see one's place in the scheme of things.

Although it is not difficult to think up reasons why we should value nature, it is more difficult to put them into practice. But this effort is open to everyone. The next step is to lay the primer of ecology down, find a natural area, and enter it. Then try to sense the place fully, with all senses open to

communication. Once you begin to feel the place and begin asking questions about the source of sounds, the names of organisms, and the direction
of processes, you are on your way. Apply the concepts presented here to the
specific place. Some may need adjustment. Not all will fit. There will be
many more questions than answers. In this dynamic, flexible world, which
you cannot communicate with directly and which operates on a logic different from your own, you can now begin to appreciate the nature of being
a human being. And this sense of self will permit you to understand more
clearly the nature of dynamism, connectedness, and creativity.

Readings

Callicott, J. Baird. 1987. *Companion to "A Sand County Almanac": Interpretive and Critical Essays*. Madison, University of Wisconsin Press.

Jansson, AnnMari, Monica Hammer, Carl Folke, and Robert Costanza. 1994. *Investing in Natural Capital: The Ecological Economics Approach to Sustainability*. Covelo, Calif., Island Press.

Taylor, Paul W. 1981. "The Ethics of Respect for Nature." *Environmental Ethics* 3: 197–218.

Literature Cited

Allee, W. C., A. E. Emerson, O. Park, T. Park, and K. P. Schmidt. 1949. *Principles of Animal Ecology*. Philadelphia, W. B. Saunders.

Allen, Timothy F. H., and Thomas B. Starr. 1982. *Hierarchy: Perspectives for Ecological Complexity*. Chicago, University of Chicago Press.

Alling, Abigail, and Mark Nelson. 1994. *Life under Glass: The Inside Story of Biosphere 2*. Oracle, Ariz., Biosphere Press.

Andrewartha, H. G., and L. C. Birch. 1954. *The Distribution and Abundance of Animals*. Chicago, University of Chicago Press.

Ashby, Ross. 1956. *An Introduction to Cybernetics*. London, Chapman and Hall.

Ashton, P. S. 1964. "Ecological Studies in the Mixed Diptocarp Forests of Brunei State." *Oxford Forestry Memoirs* 25: 1–75.

Axelrod, R., and W. D. Hamilton. 1981. "The Evolution of Cooperation." *Science* 211: 1390–96.

Axelrod, R., 1984. *The Evolution of Cooperation*. New York, Basic Books.

Bailey, Robert G. 1995. *Description of the Ecoregions of the United States*. 2d ed. Washington, D.C., U.S. Forest Service.

Baladin, R. K. 1982. *Vladimir Vernadsky: Outstanding Soviet Scientists*. Moscow, Mir.

Baudry, Jacques. 1991. "Ecological Consequences of Grazing Extensification and Land Abandonment: Role of Interactions between Environment, Society and Techniques." *Optiones Méditerranéennes, Serie Seminaires* 15: 13–19.

Baugher, Joseph F. 1988. *The Space-age Solar System*. New York, Wiley, pp. 135–38.

Bergerud, A. T. 1980. "A Review of the Population Dynamics of Caribou and Wild Reindeer in North America," pp. 556–81. In *Proceedings of the 2nd International Reindeer/Caribou Symposium*. Roros, Norway.

Berry, R. J. 1978. "Genetic Variation in Wild House Mice: Where Natural Selection and History Meet." *American Scientist* 66: 52–60.

Billings, W. Dwight. 1952. "The Environmental Complex in Relation to Plant Growth and Distribution." *Quarterly Review of Biology* 27, no. 3: 251–65.

Birch, L. C. 1953. "Experimental Background to the Study of the Distribution and Abundance of Insects." Part 3: "The Relations between Innate Capacity for Increase and Survival of Different Species of Beetles Living Together on the Same Food." *Evolution* 7: 136–44.

239

Blunt, Wilfrid. 1971. *The Compleat Naturalist: A Life of Linnaeus*. New York, Viking.

Briggs, John C. 1991. "A Cretaceous-Tertiary Mass Extinction?" *Bioscience* 41: 619–24.

Brody, Samuel. 1945. *Bioenergetics and Growth*. New York, Reinhold.

Caswell, H. 1989. "Life-history Strategies," pp. 285–307. In J. M. Cherrett, ed., *Ecological Concepts: The Contribution of Ecology to an Understanding of the Natural World*. Oxford, Blackwell Scientific Publishers.

Chapin, F. Stuart. 1980. "The Mineral Nutrition of Wild Plants." *Advances in Ecology and Systematics* 11: 233–60.

Christian, J. J., and D. E. Davis. 1964. "Endocrines, Behavior and Population." *Science* 146: 1550–60.

Clements, Frederic. 1916. *Plant Succession: An Analysis of the Development of Vegetation*. Washington, Carnegie Institution of Washington.

Cohen, Joel E., and David Tilman. 1996. "Biosphere 2 and Biodiversity: The Lessons So Far." *Science* 127: 1150–51.

Cole, Lamont. 1958. "The Ecosphere." *Scientific American* 198: 2–7.

Commoner, Barry. 1971. *The Closing Circle: Nature, Man and Technology*. New York, Knopf.

Congdon, J. D., A. E. Dunham, and D. W. Tinkle. 1982. "Energy Budgets and Life History of Reptiles," pp. 233–71. In C. Gans, ed., *Biology of the Reptilia*. New York, Academic Press.

Cook, Robert C., ed. 1962. "How Many People Have Ever Lived on Earth?" *Population Bulletin* 18, no. 1: 1–18.

Cowles, Henry. C. 1901. "The Physiographic Ecology of Chicago and Vicinity: A Study of the Origin, Development, and Classification of Plant Societies." *Botanical Gazette* 31: 73–108, 145–82.

Cronon, William. 1983. *Changes in the Land: Indians, Colonists and the Ecology of New England*. New York, Hill and Wang.

Crosby, Alfred. 1986. *Ecological Imperialism: The Biological Expansion of Europe, 900–1900*. Cambridge, Cambridge University Press.

Crumley, Carole L. P. 1987. "A Dialectical Critique of Hierarchy," pp. 155–69. In T. C. Patterson and C. W. Gailey, eds., *Power Relations and State Formulation*. Philadelphia, American Anthropological Association.

Currie, David J., and Viviane Paquin. 1987. "Large-Scale Biogeographical Patterns of Species Richness of Trees." *Nature* 329: 326–27.

Cyr, Helene, and Michael Pace. 1993. "Magnitude and Patterns of Herbivory in Aquatic and Terrestrial Ecosystems. *Nature* 361: 148–50.

Darling, F. Fraser. 1937. *A Herd of Red Deer: A Study in Animal Behavior*. London, Oxford University Press.

Deag, John M. 1980. *Social Behavior of Animals*. London, Edward Arnold.

Degens, Egon T. 1989. *Perspectives on Biogeochemistry*. Berlin, Springer.

Desy, E. A., and G. O. Batzli. 1989. "Effects of Food and Predation on Density of Prairie Voles: A Field Experiment." *Ecology* 70: 411–21.

Dugatkin, L. A., and M. Alfieri. 1991. "Guppies and the Tit for Tat Strategy: Preference Based on Past Interaction." *Behavioral Ecology and Sociobiology* 28: 243–47.

Ehrlich, Paul R., and Peter H. Raven. 1964. "Butterflies and Plants: A Study in Coevolution." *Evolution* 18: 586–608.

Ehrlich, Paul R., and Dennis D. Murphy. 1987. "Conservation Lessons from Long-term Studies of Checkerspot Butterflies." *Conservation Biology* 1, no. 2: 122–31.

Eitel, Ernest J. 1984. *Feng-shui: The Science of Sacred Landscape in Old China*. London, Synergetic Press.

Elton, Charles. 1927. *Animal Ecology*. New York, Macmillan.

Elton, Charles, and M. Nicholson. 1942. "The Ten-Year Cycle in Numbers of Lynx in Canada." *Journal of Animal Ecology* 11, no. 2: 215–44.

Emanuel, William R., Herman H. Shuggart, and Mary P. Stevenson. 1985. "Climatic Change and the Broad-scale Distribution of Terrestrial Ecosystem Complexes." *Climatic Change* 7: 29–43.

Errington, Paul. 1934. "Vulnerability of Bob-white Populations to Predation." *Ecology* 15: 110–27.

Eyres, S. R. 1971. *World Vegetation Types*. New York, Columbia University Press.

Ferré, Frederick. 1988. *Philosophy of Technology*. Englewood Cliffs, N.J., Prentice Hall.

Forman, Richard T. 1995. *Land Mosaics: The Ecology of Landscapes and Regions*. New York, Cambridge University Press.

Forrester, J. W. 1971. *World Dynamics*. Cambridge, Mass., Wright-Allen.

Fortesque, John A. C. 1980. *Environmental Geochemistry: A Holistic Approach*. New York, Springer.

Franklin, Jerry F., and Richard T. T. Forman. 1987. "Creating Landscape Patterns by Forest Cutting: Ecological Consequences and Principles." *Landscape Ecology* 1, no. 1: 5–18.

French, N., R. K. Steinhorst, and D. M. Swift. 1979. "Grassland Biomass Trophic Pyramids," pp. 59–87. In N. French, ed., *Perspectives in Grassland Ecology*. New York, Springer.

Fyfe, W. S. 1996. "The Biosphere Is Going Deep." *Science* 273: 448.

Gates, David M. 1962. *Energy Exchange in the Biosphere*. New York, Harper and Row.

Gause, G. F. 1932. "Experimental Studies on the Struggle for Existence." *Journal of Experimental Biology* 9: 389–402.

———. 1934. *The Struggle for Existence*. New York, Macmillan.

Glacken, Clarence J. 1967. *Traces on the Rhodian Shore: Nature and Culture in Western Thought from Ancient Times to the End of the Eighteenth Century*. Berkeley and Los Angeles, University of California Press.

Gleason, H. T. 1927. "The Individualistic Concept of the Plant Association." *Bulletin of the Torrey Botanical Club* 53: 7–26.

Golley, F. B. 1960. "Energy Dynamics of a Food Chain of an Old-field Community." *Ecological Monographs* 30: 187–206.

———. 1965. "Structure and Function of an Old-field Broomsedge Community." *Ecological Monographs* 35: 113–31.

———. 1986. "Environmental Ethics and Extraterrestrial Ecosystems," pp. 211–26. In E. C. Hargrove, ed., *Beyond Space Ship Earth: Environmental Ethics and the Solar System*. San Francisco, Sierra Club Books.

———. 1994. *A History of the Ecosystem Concept: More Than the Sum of the Parts*. New Haven, Yale University Press.

Golley, F. B., J. T. McGinnis, R. G. Clements, G. I. Child, and M. J. Duever. 1975. *Mineral Cycling in a Tropical Moist Forest Ecosystem*. Athens, Ga., University of Georgia Press.

Golley, Frank B., John Pinder, P. J. Smallidge, and N. J. Lambert. 1994. "Limited Invasion and Reproduction of Loblolly Pines in a Large South Carolina Old Field." *Oikos* 69: 21–27.

Grant, P. R., and B. R. Grant. 1992. "Hybridization of Bird Species." *Science* 256: 193–97.

Grime, J. P. 1977. "Evidence for the Existence of Three Primary Strategies in Plants and Its Relevance to Ecological and Evolutionary Theory." *American Naturalist* 111: 1169–94.

Grodzinski, W., and B. A. Wunder. 1975. "Ecological Energetics of Small Mammals," pp. 173–204. In F. B. Golley, K. Petrusewicz, and L. Ryszkowski, eds., *Small Mammals: Their Productivity and Population Dynamics*. Cambridge, Cambridge University Press.

Hamazaki, Toshihide. 1995. "A Study of Matrix-Patch-Organism Interactions." Ph.D. diss., University of Georgia.

Hardin, Garrett. 1968. "The Tragedy of the Commons." *Science* 162: 1243–48.

Herrera, R., C. F. Jordan, H. Klinge, and E. Medina. 1978. "Amazon Ecosystems: Their Structure and Functioning, with Particular Emphasis on Nutrients. *Interciencia* 3, no. 4: 223–32.

Holling, C. S. 1992. "Cross-Scale Morphology, Geometry, and Dynamics of Ecosystems." *Ecological Monographs* 62, no. 4: 447–502.

Hutchinson, G. E. 1958. "Concluding Remarks." *Cold Spring Harbor Symposium on Quantitative Biology* 22: 415–27.

Jansen, Daniel H. 1966. "Coevolution of Mutualism between Ants and Acacias in Central America." *Evolution* 20: 249–75.

Jansson, Ann-Marie, and James Zucchetto. 1978. "Energy, Economic and Ecological Relationships for Gotland, Sweden: A Regional Systems Study." *Ecological Bulletin* 28. Stockholm, Swedish Natural Science Research Council.

Karr, James R. 1975. "Production, Energy Pathways, and Community Diversity in Forest Birds," pp. 161–76. In F. B. Golley and E. Medina, eds., *Tropical Ecological Systems: Trends in Terrestrial and Aquatic Research*. New York, Springer.

Keith, Lloyd. 1963. *Wildlife's Ten-Year Cycle*. Madison, University of Wisconsin Press.

Kesner, Bryon T., and Vernon Meentemeyer. 1989. "A Regional Analysis of Total Nitrogen in an Agricultural Landscape." *Landscape Ecology* 2, no. 3: 151–63.

Kira, T., and K. Yoda. 1989. "Vertical Stratification in Microclimate," pp. 55–71. In H. Lieth and M. J. A. Werger, eds., *Tropical Rain Forest Ecosystems*. Amsterdam, Elsevier.

Kirchner, James W., and John Harte. 1988. "The Gaia Hypotheses: Are They Testable? Are They Useful?" *Chapman Conference on Gaia Hypothesis Program*. San Diego, American Geophysical Union.

Kramer, P. J., and J. P. Decker. 1944. "Relation between Light Intensity and Rate of Photosynthesis of Loblolly Pine and Certain Hardwoods." *Plant Physiology* 19: 350–58.

Krebs, Charles J. 1985. *Ecology: The Experimental Analysis of Distribution and Abundance*. 3d ed. Cambridge, Harper and Row .

Kropotkin, Peter. 1902 (rpt. 1955). *Mutual Aid: A Factor of Evolution*. Boston, Extending Horizon Books.

LaChapelle, Dolores. 1988. *Sacred Land, Sacred Sex, Rapture of the Deep; Concerning Deep Ecology and Celebrating Life*. Silverton, Colo., Finn Hill Arts.

Lack, David. 1947–1948. "The Significance of Clutch Size." *Ibis* 89: 302–52, 90:25–45.

Lamotte, M. 1975. "The Structure and Function of a Tropical Savanna Ecosystem," pp. 179–222. In F. B. Golley and E. Medina, eds., *Tropical Ecological Systems: Trends in Terrestrial and Aquatic Research*. New York, Springer.

Leigh, Egbert G., Jr. 1991. "Genes, Bees and Ecosystems: The Evolution of a Common Interest among Individuals." *Trends in Ecology and Evolution* 6, no. 8: 257–62.

Leopold, Aldo. 1933. *Game Management*. Madison, University of Wisconsin Press.

Lieth, Helmut. 1975. "Primary Production of the Major Vegetation Units of the World," pp. 203–15. In Lieth and Whittaker, *Primary Productivity of the Biosphere*.

Lieth, Helmut, and Robert Whittaker. 1975. *Primary Productivity of the Biosphere*. New York, Springer.

Likens, G. E., F. H. Bormann, R. S. Pierce, J. S. Eaton, and N. M. Johnson. 1977. *Biogeochemistry of a Forested Ecosystem*. New York, Springer.

Linton, D. L. 1965. "The Geography of Energy." *Geography* 50: 197–228.

Lovelock, James E., and Lynn Margulis. 1974. "Atmospheric Homeostasis by and for the Biosphere: The Gaia Hypothesis." *Tellus* 26: 1–10.

MacArthur, R. H. 1958. "Population Ecology of Some Warblers of Northeastern Coniferous Forests." *Ecology* 39: 599–619.

MacLulich, D. A. 1957. "The Place of Chance in Population Processes." *Journal of Wildlife Management* 21: 293–99.

MacArthur, R. H., and E. O. Wilson. 1967. *The Theory of Island Biogeography*. Princeton, N.J., Princeton University Press.

MacMahon, James A. 1980. "Ecosystems over Time: Succession and Other Types of Change," pp. 27–58. In R. H. Waring, ed., *Forests: Fresh Perspectives from Ecosystem Analysis*. Proceedings of the 40th Annual Biology Colloquium. Corvallis, Oregon State University.

MacMahon, J. A., D. A. Phillips, J. V. Robinson, and D. J. Schimpf. 1978. "Levels of Organization: An Organism-Centered Approach." *Bioscience* 11: 700–704.

Margulis, Lynn, and G. Hinkle. 1988. "Biota and Gaia." *Chapman Conference on Gaia Hypothesis Program*. San Diego, American Geophysical Union.

Mayr, E. 1942. *Systematics and the Origin of Species*. New York, Columbia University Press.

Mitton, Jeffry B., and Michael C. Grant. 1996. "Genetic Variation and the Natural History of Quaking Aspen." *Bioscience* 46: 25–31.

Miyawaki, Akira, ed. 1980–1989. *Vegetation of Japan*. 10 vols. Tokyo, Shibundo (Japanese, with German or English and German summaries).

Mobius, Karl. 1877. *Die Auster und dei Austernwirthschaft*. Berlin, Wiegandt, Hempel, and Parey. Trans. H. J. Rice in *Fish and Fisheries: Annual Report of the Commission for the Year 1880*, vol. 3, no. 29, part H., pp. 681–747.

Moffat, A. S. 1996. "Biodiversity Is a Boon to Ecosystems, not Species." *Science* 271: 1497.

Myers, Norman. 1979. *The Sinking Ark: A New Look at the Problem of Disappearing Species*. Oxford, Pergamon.

Naess, Arne. 1989. *Ecology, Community and Life Style*. Ed. and trans. David Rothenberg. Cambridge, Cambridge University Press.

Neilson, Ronald P. 1987. "Biotic Regionalization and Climate Controls in Western America." *Vegetatio* 70: 135–47.

Ngwa, Fred. 1992. *An Assessment of Conflicting Value Systems Surrounding African Wildlife*. Athens, Ga., Environmental Ethics Certificate Program, University of Georgia.

Noy-Meir, Imanuel, and Eddy van der Maarel. 1987. "Relations between Community Theory and Community Analysis in Vegetation Science: Some Historical Perspectives." *Vegetatio* 69: 5–15.

Odum, E. P. 1992. "When to Confront and When to Cooperate." *INTECOL Bulletin* 20: 21–23.

Odum, H. T. 1983. *Systems Ecology: An Introduction*. New York, Wiley-Interscience.

Olson, Sigrud. 1969. *Open Horizons*. New York, Knopf.

Oosting, H. J. 1948. *The Study of Plant Communities*. San Francisco, Freeman.

Orr, David W. 1992. *Ecological Literacy: Education and the Transition to a Postmodern World*. Albany, State University of New York Press.

O'Neill, Robert, D. L. DeAngelis, J. B. Waide, and T. F. H. Allen. 1986. *A Hierarchical Concept of Ecosystems*. Princeton, Princeton University Press.

Paine, Robert T. 1966. "Food Web Complexity and Species Diversity." *American Naturalist* 100: 65–75.

Park, T. 1948. "Experimental Studies of Interspecies Competition. I: Competition between populations of the flour beetles *Tribolium confusum"* and *Tribolium castaneum.*" *Ecological Monographs* 18: 265–307.

———. 1954. "Experimental Studies of Interspecies Competition. II: Temperature, Humidity and Competition in Two Species of *Tribolium.*" *Physiological Zoology* 27: 177–238.

Pattee, Howard H., ed. 1973. *Hierarchy Theory: The Challenge of Complex Systems*. New York, Braziller.

Patten, Bernard. 1985. "Further Developments toward a Theory of the Quantitative Importance of Indirect Effects of Ecosystems." *Verhandlungen der Gesellschaft für Ökologie* 1983: 271–84.

Pearl, R. 1927. "The Growth of Populations." *Quarterly Review of Biology* 2: 532–48.

Peet, R. K. 1974. "The Measurement of Species Diversity." *Annual Reviews of Ecology and Systematics* 5: 285–307.

Peters, Robert Henry. 1983. *The Ecological Implications of Body Size*. Cambridge, Cambridge University Press.

Peterson, R. O. 1977. *Wolf Ecology and Prey Relationships on Isle Royale*. U.S. Park Service Scientific Monograph Series no. 11. Washington, D.C.

Pickett, S. T. A., S. L. Collins, and J. J. Armesto. 1987. "Models, Mechanisms and Pathways of Succession." *Botanical Review* 53, no. 3: 335–71.

Pimm, Stuart L. 1982. *Food Webs*. London, Chapman and Hall.

Polunin, Nicholas, and Jacques Grinevald. 1988. "Vernadsky and Biospheral Ecology." *Environmental Conservation* 15, no. 2: 117–22.

Pomeroy, Lawrence R. 1974. "The Ocean's Food Web: A Changing Paradigm." *Bioscience* 24: 499–504.

Porter, Warren P., and David Gates. 1969. "Thermodynamic Equilibria of Animals with Environment." *Ecological Monographs* 39: 245–70.

Porter, Warren, P., J. W. Mitchell, W. A. Beckman, and C. B. DeWitt. 1973. "Behavioral Implications of Mechanistic Ecology: Thermal and Behavioral Modeling of Desert Ectotherms and Their Microenvironment." *Oecologia* (Berlin) 13: 1–54.

Postel, Sandra L., Gretchen C. Dailey, and Paul R. Ehrlich. 1996. "Human Appropriation of Renewable Fresh Water." *Science* 271: 785.

Preston, F. W. 1948. "The Commonness and Rarity of Species." *Ecology* 29: 254–83.

R. A. K. 1994. "Between Extinctions, Evolutionary Stasis." *Science* 266: 29.

Rapoport, Amos. 1977. *The Mutual Interaction of People and Their Built Environment*. Chicago, Aldine.

Reddy, Paul. 1990. "Is Mutualism Really Irrelevant to Ecology?" *Bulletin of the Ecological Society of America* 71, no. 2: 101–02.

Regan, Thomas. 1983. *The Case for Animal Rights.* Berkeley, University of California Press.

Reps, John W. 1965. *The Making of Urban American: A History of City Planning in the United States.* Princeton, Princeton University Press.

Rosenberg, Alexander. 1985. *The Structure of Biological Science.* Cambridge, Cambridge University Press.

Ruse, Michael. 1985. *Sociobiology: Sense or Nonsense.* 2d ed. Dordrecht, Reidel.

Ryszkowski, Lech, and A. Kedziora. 1987. "Impact of Agricultural Landscape Structure on Energy Flow and Water Cycling." *Landscape Ecology* 1, no. 2: 85–94.

Salisbury, F. B., and C. W. Ross. 1978. *Plant Physiology.* 2d ed. Belmont, Calif., Wadsworth.

Scheffer, V. B. 1951. "The Rise and Fall of a Reindeer Herd." *Scientific Monthly* 73: 356–62:.

Schindler, David W. 1990. "Experimental Perturbations of Whole Lakes as Tests of Hypotheses Concerning Ecosystem Structure and Function." *Oikos* 57: 25–41.

Schindler, D. W., K. H. Mills, D. F. Malley, D. L. Findlay, J. A. Sheaver, I. J. Davies, M. A. Turner, G. A. Lindsey, and D. R. Cruikshank. 1985. "Long-Term Ecosystem Stress: The Effects of Years of Experimental Acidification on a Small Lake." *Science* 228: 1395–1401.

Scholander, P. F., R. Hock, V. Walters, F. Johnson, and L. Irving. 1950. "Heat Regulation in Some Arctic and Tropical Birds and Mammals." *Biological Bulletin* 99: 237–58.

Schulze, E.-D., and H. A. Mooney, eds. 1993. *Biodiversity and Ecosystem Function.* Berlin, Springer.

Shepard, Paul. 1982. *Nature and Madness.* San Francisco, Sierra Club Books.

Signor, Philip W. 1985. "Real and Apparent Trends in Species Richness through Time," pp. 129–50. In J. W. Valentine, ed., *Phanerozoic Diversity Patterns.* Princeton, Princeton University Press.

Singer, Peter. 1990. *Animal Liberation.* 2d ed. New York, New York Review of Books Press.

Skinner, B. F. 1962. *Walden Two.* New York, Macmillan.

Slobodchikoff, C. N. 1988. *The Ecology of Social Behavior.* San Diego, Academic Press.

Slobodchikoff, C. N., and William C. Schulz. 1988. "Cooperation, Aggression and the Evolution of Social Behavior," pp. 13–32. In Slobodchikoff, *Ecology of Social Behavior.*

Snyder, Gary. 1990. *The Practice of the Wild.* San Francisco, North Point Press.

Solbrig, Otto T., ed. 1992. *From Genes to Ecosystems: A Research Agenda for Biodiversity.* Report of IUBS-SCOPE-UNESCO Workshop. Paris, International Union of Biological Sciences.

Southwood, T. R. E. 1978. *Ecological Methods.* London, Chapman and Hall.

Stanton, N. L. 1988. "The Underground in Grasslands." *Advances in Ecology and Systematics* 19: 573–89.

Stearns, Stephen C. 1976. "Life History Tactics: A Review of the Ideas." *Quarterly Review of Biology* 51, no. 1: 2–47.

Strong, Donald R., Jr. 1983. "Natural Variability and the Manifold Mechanisms of Ecological Communities." *American Naturalist* 122: 636–60.

Summerhays V. S., and C. S. Elton. 1923. "Contributions to the Ecology of Spitzbergen and Bear Island." *Journal of Ecology* 11: 214–86.

Swank, Wayne T., and D. A. Crossley, Jr. 1988. *Forest Hydrology and Ecology at Coweeta.* New York, Springer.

Tansley, Arthur G. 1935. "The Use and Abuse of Vegetation Concepts and Terms." *Ecology* 16, no. 3: 284–307.

Taylor, Paul. 1986. *Respect for Nature: A Theory for Environmental Ethics.* Princeton, Princeton University Press.

Thorne, James, and D. A. Pitz. 1988. "Landscape Akido: Management of Succession in Ecological Planning and Site Design." *Proceedings of the 1988 CELA Conference,* pp. 177–87. East Lansing, Mich., Council of Educators in Landscape Architecture.

Tobey, Ronald C. 1981. *Saving the Prairie: The Life Cycle of the Founding School of American Plant Ecology, 1895–1955.* Berkeley and Los Angeles, University of California Press.

Tuchman, Barbara. 1984. *The March of Folly: From Troy to Viet Nam.* New York, Knopf.

Turner, Monica G., and C. Lynn Ruscher. 1988. "Changes in Landscape Pattern in Georgia." *Landscape Ecology* 1: 241–51.

Turner, Monica, and William H. Romme. 1994. "Landscape Dynamics in Crown Fire Ecosystems." *Landscape Ecology* 9: 59–77.

Udvardy, M. D. F. 1975. *A Classification of the Biogeographical Provinces of the World.* Occasional Paper 18, International Union for Conservation of Nature and Natural Resources. Morges, Switzerland

Vannote, R. L., G. W. Minshall, K. W. Cummins, J. R. Sedell, and C. E. Cushing. 1980. "The River Continuum Concept." *Canadian Journal of Fisheries and Aquatic Science* 37: 130–37.

Verhulst, Pierre-François. 1845. "Recherches mathematiques sur le loi d'accroisement de la population." *Nouveaux mémoires de l'Académie Royale des Sciences et Belles-Lettres de Bruxelles,* ser. 2, 18:3–38.

Vernadsky, Vladimir I. 1986. *The Biosphere.* Abridged version of the French edition of 1929. Oracle, Ariz., Synergetic Press.

———. 1945. "The Biosphere and the Noosphere." *American Scientist* 33, no. 1: 1–12.

Vitousek, P. M., P. R. Ehrlich, A. H. Erhlich, and P. A. Matson. 1986. "Human Appropriation of the Products of Photosynthesis." *Bioscience* 36: 368–73.

Walter, H. 1973. *Vegetation of the Earth.* New York, Springer.

Walter, H., and H. Lieth. 1960–67. *Klima-diagram Weltatlas.* Jena, V. B. Gustav Fischer.

Walter, H., and E. Box. 1976. "Global Classification of Terrestrial Ecosystems." *Vegetatio* 32:5–81.

Westra, Laura. 1994. *An Environmental Proposal for Ethics: The Principle of Integrity.* Lanham, Md., Rowman and Littlefield.

Whittaker, R. H. 1966. "Forest Dimensions and Production in the Great Smoky Mountains." *Ecology* 47: 103–21.

Whittaker R. H., and W. A. Niering. 1975. "Vegetation of the Santa Catalina Mountains, Arizona. V: Biomass, Production and Diversity along an Elevation Gradient." *Ecology* 56: 771–90.

Williams, George. 1992. "*Gaia,* Nature Worship and Biocentric Fallacies." *Quarterly Review of Biology* 67, no. 4: 479–86.

Wilson, E. O. 1975. *Sociobiology: The New Synthesis.* Cambridge, Harvard University Press.

Wilson, E. O., and F. M. Peter. 1988. *Biodiversity.* Washington, National Academy of Science Press.

Young, Gerald L., ed. 1983. *Origins of Human Ecology.* Stroudsburg, Pa., Hutchinson and Ross.

———. 1986. "Environment: Term and Concept in the Social Sciences." *Social Science Age* 25: 83–124.

Index